ROLE OF SEDIMENT IN THE DESIGN AND MANAGEMENT OF IRRIGATION CANALS

Sunsari Morang Irrigation Scheme, Nepal

Promoter: Prof. E. Schultz, PhD, MSc
Professor of Land and Water Development
UNESCO-IHE Institute for Water Education
Delft, the Netherlands.

Co-promoter: Prof. N.M. Shakya, PhD, MSc
Professor of Water Resources Engineering
Civil Engineering Department
Institute of Engineering, Pulchowk Campus
Tribhuwan University
Kathmandu, Nepal.

Awarding committee: Prof. dr. Linden Vincent
Prof. of Irrigation and Water Engineering
Wageningen University
Wageningen, the Netherlands

Dr. ir. A.J.F. Hoitink
Assistant professor of Environmental Fluid Mechanics
Group Hydrology and Qualitative Water Management
Wageningen University
Wageningen, the Netherlands

Prof. N. Wright, PhD
Professor of Hydraulic Engineering and River Basin
Development
UNESCO-IHE Institute for Water Education
Delft, the Netherlands

Dr. Z.B. Wang
Expert Advisor
Deltares
Delft, the Netherlands

ROLE OF SEDIMENT IN THE DESIGN AND MANAGEMENT OF IRRIGATION CANALS

Sunsari Morang Irrigation Scheme, Nepal

DISSERTATION

Submitted in fulfilment of the requirements of
the Academic Board of Wageningen University
and the Academic Board of the UNESCO-IHE Institute for Water Education
For the Degree of DOCTOR
To be defended in public on
Wednesday, 17 November 2010 at 10:00 hrs
in Delft, the Netherlands

by

Krishna Prasad PAUDEL

Born in Chitwan, Nepal
Master of Science, UNESCO-IHE, the Netherlands

CRC Press/Balkema is an imprint of the Taylor & Francis Group, an informa business

© 2010, Krishna P. Paudel

Published by:
CRC Press/Balkema
PO Box 447, 2300 AK Leiden, the Netherlands
e-mail: Pub.NL@taylorandfrancis.com
www.crcpress.com – www.taylorandfrancis.co.uk – www.ba.balkema.nl
ISBN 978-0-415-61579-2 (Taylor & Francis Group)
ISBN 978-90-8585-851-5 (Wageningen University)

Table of contents

Acknowledgement

During the long-lasting and often interrupted evolution of this research I have accumulated many debts, only a portion of which I have space to acknowledge here. I am grateful to many people for their help, both direct and indirect, during this research. Foremost, I would like to express profound gratitude to my promoter Prof. Bart Schultz for his invaluable support, encouragement, supervision and useful suggestions throughout this research work. His moral support and continuous guidance enabled me to complete my work successfully.

My sincere thanks are to my mentor Mr. Herman Depeweg, who has been actively interested in my work and has always been available to advise me. I am very grateful for his patience, motivation and enthusiasm. His immense knowledge in hydraulics and irrigation became instrumental in giving this thesis the final shape. I am also highly thankful to Dr. Suryadi for his suggestions, comments and cooperation.

The study would not have been possible without the financial support from the Netherlands Organization for International Cooperation in Higher Education (NUFFIC). I express my sincere gratitude. I would also like to thank all the academic and administrative staff of UNESCO-IHE for providing the friendly working atmosphere especially, Jolanda Boots and Tonneke Morgenstond for their continuous support and understanding that helped me to overcome all the difficulties of being a sandwich student and made me feel very much a part of the Institute. They tirelessly responded to my numerous administrative requests.

I would like to acknowledge the support and favour extended by all the staff of Sunsari Morang Irrigation Scheme. Special thanks are to Mr. Anil Kumar Pokharel, my colleagues, field staff and members of the Water Users Association, who provided all the possible cooperation during the fieldwork. Above all, I am grateful to Mr. Ram Padarath Shah without whom, the fieldwork would not have been possible. He has immense information in the field of sediment. I am also thankful to Sanjeev Mishra and Kushang Sherpa for their support during the data collection and processing.

I am fortunate to have many friends particularly, Ashok Chapagain, and Sarfraz Munir. I am thankful for their moral support and company. My special thanks are due to Gerard Pichel (DHV) for his encouragements and valuable tips in analyzing sediment transport problems. Further, thanks should also go to Jalbert and Jitka for their help to provide me a peaceful working environment during my stay at Lochem. Last but not the least my thanks go to my family and friends, whose support and affection encouraged me to complete my research successfully. Finally most of all to my wife Saraswati and our children Rabin and Rosha for the endless patience and understanding they showed during my absence.

Summary

Sediment transport in irrigation canals

The sediment transport aspect is a major factor in irrigation development as it determines to a large extent the sustainability of an irrigation scheme, particularly in case of unlined canals in alluvial soils. Investigations in this respect started since Kennedy published his channel-forming discharge theory in 1895. Subsequently different theories have been developed and are used around the world. All of them assume uniform and steady flow conditions and try to find the canal dimensions that are stable for a given discharge and sediment load. In the past irrigation schemes were designed for protective purposes with very little flow control, hence steady and uniform flow conditions could be realised to some extent.

Modern irrigation schemes are increasingly demand based, which means that the water flow in a canal is determined by the crop water requirements. Accordingly the flow in the canal network is not constant as the crop water requirement changes with the climate and the growing stages of the crops. Also the inflow of the sediment is not constant throughout the irrigation season in most schemes. The situation is even worse for run-of-the-river schemes where fluctuations in the river discharge have a direct effect on the inflow of water and sediment.

The conventional design methods are not able to predict accurately the sediment transport behaviour in a canal, firstly due to the unsteady and non-uniform water flow conditions and secondly due to the changing nature of the sediment inflow. Hence, the actual behaviour of a canal widely diverges from the design assumptions and in many cases immense maintenance costs have to be met with to tackle the sediment problems.

An irrigation scheme should not only be able to deliver water in the required amount, time and level to the crops on the field, but also should recover at least its operation and maintenance cost. Cost recovery is, to some extent, related to the level of service provided by the irrigation organization and the expenditure for operation and maintenance of the scheme. Past experiences in Nepal have shown that modernization of existing irrigation schemes to improve the level of service has also increased the operation and maintenance costs. These costs are, in some cases, high compared to the generally low level of ability of the water users and farmers to pay these costs. The search of making schemes more equitable, reliable and flexible has resulted in the introduction of new flow control systems and water delivery schedules that may, if not carefully designed, adversely affect the sediment transport behaviour of a canal. In quite some schemes unpredicted deposition and/or erosion in canals have not only increased the operation and maintenance costs but also reduced the reliability of the services delivered.

Irrigation development in Nepal and the study area

Nepal is a landlocked country in South Asia lying between China and India. It is situated between 26°22' N to 30°27' N latitude and 80°4' E to 88°12' E longitude of the prime meridian. Roughly rectangular in shape, the country has an area of 147,181 km^2. It is 885 km in length but its width is uneven and increases towards the West. The mean North-South width is 193 km. Nepal is a predominantly mountainous country, with elevations ranging from 64 m+MSL (Mean Sea Level) at

Kechana, Jhapa to 8,848 m+MSL at the peak of the world highest mountain, Everest, within a span of 200 km. Nepal has a cultivated area of 2.64 million ha, of which two third (1.77 million ha) is potentially irrigable. At present 42% of the cultivated area has some sort of irrigation, out of which only 41% is receiving year round irrigation water. The existing irrigation schemes contribute approximately 65% of the country's current agriculture production.

Nepal has a long history of irrigated agriculture. Most of the existing large-scale irrigation schemes are located in the southern alluvial plain (Terai). The canals are unlined and the sediment load forms an integral part of the supplied irrigation water. The schemes are predominantly supply based and have a very low duty for intensive cropping. In view of the increased competition among the different water using sectors and low performance of these schemes, many of them are undergoing modernization. For example, the Sunsari Morang Irrigation Scheme (SMIS) is one of the schemes under modernization, and it has been taken as a study case for this research. A better understanding of the sediment transport process under changing flow and sediment load conditions, a shifting management environment and different maintenance scenarios will be very useful in pulling out the schemes from the present vicious cycle of construction-deterioration-rehabilitation.

The Sunsari Morang Irrigation Scheme (SMIS) is located in the eastern Terai. The Koshi River is the source of water. A side intake for the water diversion, an around 50 km long main canal of capacity 45.3 m^3/s for water conveyance and 10 secondary canals and other minor canals of various capacities for water distribution were constructed to irrigate a command area of 68,000 ha. The system was put into the operation in 1975, but faced a serious problem of water diversion and sediment deposition in the canal network. Hence from 1978, after 3 years of operation, rehabilitation and modernization work of the scheme has been started. During modernization the intake has been relocated to increase the water diversion and reduce the sediment entry. Besides, a settling basin with dredgers for continuous removal of sediment has been provided near the head of the main canal. Apart from that the command area development and modernization of existing canal network is in progress and till third phase (1997-2002), around 41,000 ha area has been developed.

Sediment transport research

The aim of this research is to understand the relevant aspects of sediment transport in irrigation canals and to formulate a design and management approach for irrigation schemes in Nepal in view of sediment transport. In the process, the design methods used in the design of irrigation schemes in Nepal and their effectiveness on sediment transport have been studied. The impact of operation and maintenance on sediment movement has been analysed taking the case study of SMIS. An improved design approach for sediment transport in irrigation canals has been proposed. A mathematical model SETRIC has been used to study the interrelationship of sediment movement with the design and management and to evaluate the proposed design approach for irrigation canal based on the data of the SMIS.

The mathematical formulation of sediment transport process in an irrigation canal is based on the previous works in this field, most notably the work of Mendez on the formulation of the mathematical model SETRIC. Subsequent analysis, improvement and verification works by Paudel, Ghimire, Orellana V., Via Giglio and Sherpa have been used. The model SETRIC has been verified and improved where found necessary and has been used to analyse the irrigation scheme and to propose an improvement in the design and management from sediment transport point of view.

Assessment of design parameters

The methods of selecting the design discharge and sizing of canals for modern irrigation schemes based upon the present concept of crop based irrigation demand, water delivery schedules and water allocation to the tertiary units have been analysed. The selection of a crop depends upon the soil type, water availability, socio-economic setting and climatic conditions. The type of crop together with the soil type determines the irrigation method and irrigation schedules, while the type of crop and climatic condition determines the irrigation water requirement. The required flow in a canal is then derived based on the water delivery schedule from that canal to the lower order canals or to the field to meet the water requirement.

The factors that influence the roughness of an irrigation canal have been analysed and a proposal for a more rational roughness determination process has been formulated based on the available knowledge. The roughness in the sides depends upon the shape and size of material, vegetation and surface irregularities, while the roughness in the bed is a function of shape and size of material and the surface irregularities (bed form in case of alluvial canals). For the prediction of roughness in the bed mostly two approaches are in use – methods based on hydraulic parameters (water depth, flow velocity and bed material size) and the methods based on bed forms and the grain related parameters. In this research, the method based on the bed form and grain related parameters, as suggested by van Rijn, has been used. Similarly, for the determination of roughness in the sides, the influence of surface irregularities have been included by dividing the maintenance condition as ideal, good, fair and poor and accordingly applying the correction to the standard roughness value for the type of material. The influence of vegetation has been accounted based on the concept of V.T. Chow. The various methods of computing the equivalent roughness have been compared and the method proposed by Mendez has been found to be better when tested with the Kruger data.

Most of the sediment transport predictors consider the canal with an infinite width without taking into account the effects of the side walls on the water flow and the sediment transport. The effect of the side wall on the velocity distribution in lateral direction is neglected and therefore the velocity distribution and the sediment transport are considered to be constant in any point of the cross section. Under that assumption a uniformly distributed shear stress on the bottom and an identical velocity distribution and sediment transport is considered. Majority of the irrigation canals are non-wide and trapezoidal in shape with the exception of small and lined canals that may be rectangular. In a trapezoidal section the water depth changes from point to point in the section and hence the shear stress. The effect would be more pronounced if the bed width to water depth ratio (B-h ratio) is small. The change in velocity distribution in a canal in view of the change in boundary shear and water depth along the cross section has been analysed and evaluated with the field measurements. The change in velocity and shear stress in a canal section has been used to evaluate the influence of B-h ratio and side slope in the prediction of sediment transport capacity by selected predictors (Brownlie, Engelund-Hansen and Ackers-White). The evaluation with the available data set showed that the proposed correction improved the predictability for non-wide irrigation canals.

Canal design approaches for sediment transport in Nepal

For the design of canals having erodible boundary and carrying sediment loads two approaches are in practice, namely the regime method and the rational method. The regime design methods are sets of empirical equations based on observations of canals and rivers that have achieved dynamic stability. The rational methods are more analytical in which three equations, an alluvial resistance relation, a sediment transport equation and a width-depth relationship, are used to determine the slope, depth and width of an alluvial canal when the water and sediment discharges as well as the bed material size are specified.

In Nepal, the design manuals of the Department of Irrigation recommend Lacey's regime equations and White-Bettess-Paris tables with the tractive force equations for the design of earthen canals carrying sediment. But in practice, there is no consistency in the design approaches that has been found to vary from canal to canal even within the same irrigation scheme. The use of Lacey's equation for computing the B-h ratio has generally resulted in wider canals. This is so, because flatter side slopes than predicted by the Lacey's equations are used from soil stability considerations.

The White-Bettess-Paris tables are derived from alluvial friction equations of White, Bettess and Paris (1980) and sediment transport equations of Ackers and White (1973). No records regarding the use of this method for the design of canals was found and hence its performance in terms of sediment transport could not be verified. However, the Ackers and White sediment transport equations over-predicted the sediment transport capacity of a canal when tested with the SMIS data. The sediment load entering into the canals of SMIS is mostly fine ($d_{50} < 0.2$ mm) and most of the large scale irrigation schemes in Nepal have similar geo-morphological settings. That means that the White-Bettess-Paris tables will result in a canal with a flatter slope than actually required to carry the type of sediment prevailing in SMIS and other similar irrigation schemes of Nepal. Analysis showed that the Brownlie and Engelund and Hansen equations are more suitable for the type of sediment that has been found in SMIS.

During the modernization, the secondary canals (S9 and S14) of SMIS have been designed by two different approaches. Secondary Canal S9 has been designed using Lacey's regime concept while Secondary Canal S14 has been designed using an energy approach. In the energy approach the erosion is controlled by limiting the tractive force and the deposition is controlled by ensuring equal or non-decreasing energy of the flow in the downstream direction. Both the canals have been evaluated for their sediment transport capacity for the prevailing sediment characteristics. The carrying capacities of both canals (~ 230 ppm) have been found to be less than the expected sediment load (~ 300 – 500 ppm) in the canal. The energy concept assumes that the sediment transport is proportional to the product of velocity and bed slope. The carrying capacity of the canal designed by this principle has been found to be variable along its length. It means that the sediment transport capacity is not only a function of bed slope and water depth as assumed in the energy concept.

An improved approach for the design and management of irrigation canals

In general the reliability of sediment transport predictors is low and at best they can provide only estimates. As per Vito A Vanoni (1975) a probable error in the range of 50-100% can be expected even under the most favourable circumstances. There is no universally accepted formula for the prediction of sediment transport. Most of them are based upon laboratory data of limited sediment and water flow ranges. Hence they should be adjusted to make them compatible to specific purposes, otherwise the predicted results will be unrealistic. An improved rational approach has been proposed for the design of alluvial canals carrying sediment loads. To find the bed width, bed slope and water depth of a canal for a given discharge and sediment characteristics three equations, namely a sediment transport predictor (total load), resistance equation (Chézy) and a B-h ratio predictor are used.

A canal design program DOCSET (Design Of Canal for SEdiment Transport) has been prepared for the improved approach including the above mentioned improvements. The program can also be used to evaluate the existing design for a given water flow and sediment characteristics. Basic features of the new approach are:

– *concept of dominant concentration*. Instead of using the maximum concentration, the approach suggests to look for a concentration that results in net minimum erosion/deposition in one crop calendar year;

– *determination of roughness*. The proposed method makes use of the elaborated and more realistically determined roughness value in the design process. The roughness of the cross section is adjusted as per the hydraulic condition and sediment characteristics. Moreover the influences of the side slopes and the B-h ratio are included while computing the equivalent roughness of the section. This should result in a more accurate prediction of hydraulic and sediment transport characteristics of the canal and hence, a better design;

– *explicit use of sediment parameters*. The sediment concentration and representative size (d_m) is explicitly used in the design. That will make the design process more flexible as different canals might have to divert and convey sediment loads of different sizes (d_m) and amounts;

– *use of an adjustment parameter*. An adjustment parameter has been used that includes the influence of non-wide canals, sloping side walls and exponent of velocity in the sediment transport predictor. This adjustment should increase the accuracy of the predictors when they are used in irrigation canals, an environment for which they were not derived;

– *holistic design concept*. This approach uses one canal system as a single unit. The canal system may have different canals of different levels, but the water and sediment management plans are prepared for the whole system. Then the hydraulic design of the individual canal can be made to meet the design management plan for that canal;

– *selection of B-h ratio*. A B-h ratio selection criterion has been proposed considering the side slope selection practices in Nepal as well as the sediment transport aspects.

Since, the sediment transport process is influenced by the management of the irrigation scheme, the design should focus to have a canal that is flexible enough to meet the demand and still have a minimum deposition/erosion. The provision of sufficient carrying capacity up to the desired location (conveyance), providing controlled deposition options if the water delivery plans limit the transport capacity (provisions of settling pockets) and preparation of maintenance plans (desilting works) are some of the aspects that would have to be analysed and included in the design to reduce the sediment transport problems.

The canal design methods can give the best possible canal geometry for a given water flow and sediment concentration only. For water flows and sediment concentrations other than the design values, there may be either erosion or deposition. The aim of the design would have to be to balance the total erosion and deposition in one crop calendar year. So, a design may not be based on the maximum sediment concentration expected during the irrigation season, but on a value that results in the minimum net erosion/deposition. The best way to evaluate a canal under such scenario is to use a suitable sediment transport model. Besides, the roughness of the canal depends upon the hydraulic conditions, sediment characteristics and the maintenance plans that are constantly changing throughout the irrigation season. The canals are designed assuming a uniform flow and sediment transport under equilibrium condition. However, such conditions are seldom found in irrigation canals due to the control in flow to meet the variation in water demand. Hence, the design of a canal would have to be evaluated using a sediment transport model for the selection of proper design parameters and to evaluate the design for the proposed water operation plans.

The mathematical model SETRIC

The mathematical model SETRIC is a one-dimensional model, where the water flow in the canal has been schematised as a quasi-steady and gradually varied flow. This one dimensional flow equation is solved by the predictor-corrector method. Galappatti's depth integrated model for sediment transport has been used to predict the actual sediment concentration at any point under non-equilibrium conditions. Galappatti's model is based on the 2-D convection-diffusion equation. The mass balance equation for the total sediment transport is solved using the modified Lax's method, assuming a steady condition of the sediment concentration. For the prediction of the equilibrium concentration one of the three total load predictors: Brownlie, Engelund and Hansen or Ackers and White methods can be used.

The model SETRIC was evaluated using other hydrodynamic and sediment transport models (DUFLOW and SOBEK-RIVER) and was validated by the field data of SMIS. Predictability of different predictors has been compared. The Brownlie and Engulund and Hansen methods predicted reasonably for the sediment size of 0.1 mm (d_{50}), while predictability of Ackers and White for the sediment size was found to be poor. The sensitivity of Brownlie's method was more uniform than the other two methods for a sediment size range of 0.05 to 0.5 mm.

Field data collection

For the field measurements of the sediment transport process, one of the secondary canals of SMIS (S9) was selected. Since, the objective of field data was to test the design approach for sediment transport; preference was given for a canal that was recently designed and constructed. The field measurement of water and sediment flow was carried out in 2004 and 2005. During field measurements the water inflow rate into Secondary Canal S9 system was measured. A broad crested weir immediately downstream of the intake for Secondary Canal S9 was calibrated and used for discharge measurement. For sediment concentration measurements, dip samples just downstream of the hydraulic jump were taken on a daily basis. The samples were then analysed in the laboratory and the sediment concentration was determined. Point sampling across the section using pump samplers were also taken and the calculation results showed that the dip samples underestimated pump samples by around 8% in case of the total load and by around 35% for the sediment of size > 63 μm. At the end of the irrigation season, the deposited sediment samples along the canal were taken to determine the representative sediment size and other properties.

The irrigating water delivered to the sub-secondary canals, delivery schedules and the set-points upstream of water level regulators were also measured. For morphological change, the pre-season and post-season canal geometry was measured. The velocity distribution in the trapezoidal earthen canal section was measured. Besides, field measurement roughness (indirect measurement) was also made in the beginning, mid and end of the seasons to determine the change in roughness in time.

Modelling results

The model SETRIC was used to study the effect on the sediment transport process due to system management activities namely, change in water demand and supply, water delivery modes based on the available water and the change in sediment load due to the variation in sediment inflow from the river or problems in proper operation of the settling basin. For the design water inflow into Secondary Canal S9, a water delivery schedule has been designed and has been evaluated for sediment transport efficiency under changing sediment inflow conditions. The improved canal design approach was evaluated comparing the results with the existing design of Secondary Canal S9. Some of the findings of the modelling results are:

− water delivery schedules can be designed and implemented to reduce the erosion/deposition problems of a certain reach even after the system is constructed and put into operation;

− the design operation plans and assumptions have not been followed in Secondary Canal S9 of SMIS. From sediment transport perspective, the existing water management practice results more sedimentation in the sub-secondary and tertiary canals than Secondary Canal S9;

− the periodic change in the demand and the corresponding change in sediment transport capacity of the canal can be manipulated to arrive at the seasonal balance in the sediment deposition. In one period, there may be deposition but that can be eroded in the next period;

− the proposed water delivery plan is based on the existing canal and its control structures and covers discharge fluctuation from around 46% to 114%. Hence, it can be implemented with the present infrastructure and can handle all possible flow situations in the canal;

− the proposed delivery schedule ensures either full supply or no supply to the sub-secondary canals which have been designed for the same principle. This could reduce the existing deposition problem faced by these canals;

− proper operation of the settling basin is crucial for the sustainability of the SMIS;

– the secondary canals need to be operated in rotation when there is less demand or less available water in the main canal. This will ensure design flow in the secondary canals and reduce the sedimentation problem. The main canal would have to be analysed for the best mode of rotation from sediment transport and water delivery perspective.

Major contributions of this research

Apart from the recommendations made in the design, management and operation of Secondary Canal S9 from sediment transport perspective, the following contributions are made by this research:

– an elaborate analysis of the velocity and shear stress distribution across the trapezoidal canal is made to derive the correction factor for the sediment transport predictor. This will help to increase the predictability when the predictors are used for the analysis of sediment transport in irrigation canals;

– an explicit method of including roughness parameters in the calculation of the equivalent roughness for the mathematical model has been proposed;

– the sediment transport model SETRIC has been updated and its functionality has been improved. The model can now be used as a design as well as research tool for analyzing the sediment transport process for different water delivery schedules and control systems;

– an improved approach for the design and management of irrigation canals has been proposed. A computer program DOCSET has also been prepared based on the approach. The program is interactive, easy to use and can be used by designers with limited modelling know-how;

– a water delivery plan has been designed and tested for the changing water and sediment inflow condition that can be implemented with the existing canal infrastructure;

– the causes of sedimentation in the sub-secondary canals of Secondary Canal S9 have been identified.

Conclusions and outlook for the future

Canal design is an iterative process where the starting point is the preparation of management plans. Then the design parameters need to be selected and the preliminary hydraulic design of the canal can be made. The design results can then be used in the model to simulate and evaluate the proposed management plans and the sediment transport process in the system. Necessary adjustment can be made either in the design parameters or in the management plans, if needed. Then the canal need to be redesigned and the process would have to be continued till a satisfactory condition is reached.

The coarser fraction of the sediment is mostly controlled at the headwork and settling basin of an irrigation scheme. The sediment that is encountered in main and secondary canals is generally fine sand. Most of the silt fraction (sediment < 63 μm) is transported to the lower order canals and fields where it gets deposited. In the sub-secondary and tertiary canals, it has been observed that the fine sediment does not roll down to the bed as normally assumed in the case of sand and gets deposited on the slope also. Thus the canal section becomes narrower and the side slope becomes steeper. This phenomenon can not be analysed with the present sediment transport assumptions and an investigation in this aspect to address the transport process of fine sediment would be beneficial for improving the design and management of irrigation canals.

Flexibility of operation and sediment transport aspects restrict each other. A canal without any control can be designed and operated with higher degree of reliability in terms of sediment transport. Once the flow is controlled the sediment transport pattern of the canal is changed and the designed canal will behave differently. Hence, both flexibility and efficient sediment management are difficult to achieve at the same time. A compromise has to be made and this needs to be reflected in the design.

All the methods to transport, exclude or extract the sediment are temporary measures and just transfer the problem from one place to the other. They are not the complete solutions of the sediment problem. A better understanding on sediment movement helps to identify the problems beforehand and look for the best possible solutions.

1
Introduction

1.1   General

Ideally irrigation schemes should be able to provide water in time, amount and with desirable head to the agricultural field. The irrigation water demand keeps on changing throughout the irrigation season as it depends upon the climatic conditions, type and stage of crops and soil moisture conditions. So a canal network has to carry the variable amounts of flow, mostly less than the discharge that it was designed for. The design discharge can be defined as the maximum amount of flow that can be handled in a proper way. Various factors like crop water requirement, irrigation methods, water distribution plans, flow control mechanism and socio-economic settings are considered in determining the design discharge.

Various methods are available for the design of canals. Some use basic principles of hydraulics and soil stability to determine the geometry of the canal. Tractive force methods (Fortier and Scobey, 1926, Lane, 1955), rational methods (Chang, 1980, White, *et al.*, 1981b) are some of the methods in this category. Some methods have been evolved from the study of relatively stable canals around the world. These methods are known as regime methods and the works of Lacey (1930) and Simons and Albertson (1963) are few examples in this field. Suitable design approaches can be used depending upon whether the canal has a rigid boundary or has an erodible boundary and is carrying clear water or has an erodible boundary and is carrying water with sediment.

Canals are generally designed assuming steady and uniform flow. However, this situation is seldom found in a modern irrigation scheme. Modern irrigation schemes are increasingly demand oriented and require frequent operation of control gates that leads to unsteady and non-uniform flow. The design becomes more complicated in case the canal has an erodible boundary and carries water with sediment. Most schemes in this category require a large amount of maintenance due to unwanted deposition on or erosion of the canal bed and banks. Efficient hydrodynamic models are available to simulate the flow for different gate operation and inflow rates. These models are being extensively used to verify the hydrodynamic performance of the canal network for design and modernization purposes.

Although certain similarities exist between irrigation canals and rivers, the sediment transport models for rivers are not applicable for canals due to the specific differences between rivers and canals, among others the appropriate use of sediment transport formulae and friction factor predictors, the effect of the canal sides on the velocity distribution and sediment transport, and the operation rules. The sediment transport concepts should be related to the flow conditions and sediment characteristics prevailing in irrigation canals. Few models exist that are meant for canal networks like ISIS (Halcrow, 2003), Sediment and Hydraulic Analysis for Rehabilitation of Canals (SHARC) (HR Wallingford, 2002), Simulation of Irrigation Canals (SIC) (Malaterre and Baume, 1997), but these models do not include explicitly the effect of canal side slopes on the velocity distribution and of maintenance on the sediment movement.

With a slight change in water flow and sediment properties the sediment movement pattern may be affected significantly. Operation and maintenance of an irrigation scheme has a major influence on the hydrodynamic behaviour of the canal and hence on the sediment movement also. Analyzing the problem from a design point of view is not sufficient to solve the problem. An integrated approach that looks into the design as well as management aspects simultaneously is needed to deal with the sediment movement problems in irrigation canals.

The aim of this research is to evaluate the design concept of irrigation canals in view of water and sediment transport. With the help of field data from one of the major irrigation schemes in Nepal the impact of operation and maintenance on sediment movement has been studied. The interrelationship of sediment movement with design and management has been evaluated with the help of the mathematical model SETRIC and finally an integrated design approach has been suggested.

1.2   Prospects of irrigation

Irrigation will be the key factor in maintaining food supply to the growing population of the world. Presently around 270 million ha of irrigated land, which is approximately 18% of the total cultivated land is responsible for 40% of crop output (Schultz and Wrachien, 2002). More and more area is being cultivated to increase the production, but at the same time more area is being irreversibly lost due to degrading uses and to permanent cover changes (Meyer and Turner II, 1992). Annually approximately 14.5 million ha cultivated land has to be taken out of cultivation due to urbanization, industrialization and waterlogging or salinization problems (Schultz, 2002).

It is clear that land for cultivation cannot be expanded unlimitedly; hence productivity per unit area has to be increased. From the point of view of food production there is a common feeling that 90% of the required increase will have to be realized on existing cultivated land and 10% on newly reclaimed land (Schultz, *et al.*, 2005). Irrigation is one of the major factors to increase productivity. Most of the financially attractive irrigation schemes have already been constructed. Further increment in the irrigated area is possible if large-scale multipurpose storage and inter-basin river diversion projects are constructed. Such schemes require large investments and also have to pass through environmental and safety issues. Moreover, irrigation schemes use about 70% of the withdrawn water from the global river systems and only 50% of the withdrawn water reaches the crop (Fischer, *et al.*, 2006). The competition among the water use sub-sectors is going to increase in the future.

Despite undeniable past successes in contributing to food production, irrigation expansion lost its momentum since the 1980s due to a considerable slowdown in new investment. The rate of expansion of irrigated agriculture decreased to nearly 1% per year after 1975. High cost of irrigation development (Pereira, *et al.*, 1996) and relatively balanced situation in the supply and demand of cereals (FAO, 2007) might be the reasons of decrease in the investment in irrigation development. Pluesquellec (2002) argues that the reduction in investment of irrigation is due to the relatively poor performance of large-scale canal irrigation projects. These systems are the most difficult to manage and have yielded the lowest returns compared to their expected potential. Despite all, irrigated agriculture remains essential for future food security. Hence, construction of new schemes as well as optimal use of already acquired water in constructed schemes is equally important for the increment in the irrigated area.

1.3   Essentials of irrigation schemes

The success of a project is essentially determined by the creation of an attractive environment for the users to initiate and continue the proposed activities (Constandse, 1988). The irrigation schemes should be able to meet the expectation of the users in terms of the services and motivate them to payback for the services they use. In the design of water management systems in rural areas, the determination of the required level of service is a complicated matter, as the interaction between water management and crop yield is difficult to quantify (Schultz, 2002).

1.3.1   Functional requirement

An irrigation scheme should be able to deliver water in the required amount, time and level to the field. From the farmers' view there are three criteria of water delivery whether the supply is sufficient (adequacy), is in the right time (reliability), is a fair share compared to the other farmers (equity).

Adequacy

The capacity of a canal to deliver the design (required) discharge is the first and the most important pre-requisite of a successful scheme. This aspect might have been overlooked in the planning stage, design stage or even during the construction stage. Inappropriate selection of water delivery structures and/or defects in construction make it impossible to meet the adequacy (Murray-Rust and Halsema, 1998). Adequacy may be affected primarily due to:
−   reduced canal capacity due to sediment deposition;
−   available water is not sufficient to meet the demand;
−   canal capacity is low as compared to the command area;
−   control structures are not compatible with the mode of operation in practice.

Reliability

Reliability is an expression of confidence in the irrigation scheme to deliver water as in the design. It is defined as the ratio of the amount delivered to that scheduled. It is important to the farmers because it allows them for proper planning. By ensuring an adequate and reliable supply of water, irrigation may increase yields of most crops by 100 to 400% (Food and Agriculture Organization, 2003). Unreliable water delivery makes farmers reluctant to invest in terms of improved seed and fertilizer to increase the yields.

Delivery of water in time is influenced by the water delivery schedules, sensitivity of the off-take structures, and hydrodynamic performance of the canal network in combination with the water level regulators and maintenance condition of the schemes. The performance of the off-take is highly influenced by the upstream and downstream water level (Renault, 2003, Ankum, 2004), which is affected by sediment deposition. The ultimate purpose of fixing the canal capacity and providing flow control structures would have to meet the proposed irrigation schedules, which again is to provide water in time. Hence all the appurtenances in the scheme would have to support each other to attain the targeted objectives.

Equity

Equity can be defined as the delivery of a fair share of water to users throughout the scheme. Inequitable distribution results in the overuse of water by the head reach farmers and little use by the tail-enders. The implications are wastage of diverted water, poor coverage of the intended command area and social injustice. This is one of the major reasons of poor irrigation service fee collection and the participation of users in the operation and maintenance of the scheme. Major causes of the inequity are:

- *sediment deposition in the canals*. Removal of sediment from a canal network is an expensive activity and most of the time the allocated annual operation and maintenance budget is insufficient to remove the deposited material. The general trend is to wait for the next major rehabilitation for the sediment to be removed completely and maintain the original level and geometry. During the period, from one major rehabilitation to another, the upstream off-takes are benefited due to the raised water level while the tail end off-takes suffer from reduced canal capacity and unfair drawing of water (Belaud and Baume, 2002);
- *control structures*. Sensitivity analysis of water level and discharge regulators is seldom done during construction and/or modernization. Type and setting of off-take structures have a major role in equitable distribution of water and decides how they will perform under changing flow conditions;
- *selection of design parameters*. Selection of reasonable design roughness values, fixing the width and crest height of the conveyance (control) structures for no drawdown and backwater conditions and proper determination of head losses are some of the aspects that need to be taken care of during design to maintain equity. The tendency is to select lower roughness values to arrive at more economical sections, which have resulted in some cases in the inability of the main canal to draw the design discharge as well as inequity in distribution (Pradhan, 1996).

1.3.2   Service requirement

On demand water delivery service is the ideal level of service any farmer would like to have in his scheme. However, without the provision of reservoirs the on demand type of water delivery service is generally not possible. Moreover service requirements of irrigation schemes change with time due to an increased competition among the water use sub-sectors and pressure to meet the food demand of the growing population. The scheme should be such that the operation and maintenance requirements are to the level of farmer's skill and capacity so that they can manage the scheme in a sustainable way. A balance has to be maintained such that the level of service is attractive enough and at the same time the operation and maintenance costs are economically affordable by the farmers. Then only the scheme can be managed in a sustainable way. Construction/modernization of a scheme in a traditional way may solve the problem temporarily but the prime objective of creating a sustainable irrigation scheme would not be fulfilled.

Another consideration of determining the level of service is the possible benefit from agriculture and the socio-economic conditions of the people. In least developed countries, with the present economic conditions and type of land tenure, a high level of service is not possible in the near future. Even if the prices of agricultural products increase, the other infrastructure does not support for the commercialisation of agriculture at short notice.

1.3.3   Operation and maintenance requirement

The major problem in the development of irrigation schemes is the inability to recover at least the operation and maintenance (O&M) cost. The schemes cannot generate the required O&M funds from the irrigation service fee. This is due to high O&M cost and low rate of payback from the users. Irrigation schemes carrying sediment laden water require large amounts of money for the maintenance of the canal network. Each scheme has its typical problem that may need a large O&M budget, but in general most of the expenditure is incurred in the following components:
- *intake*. The O&M cost varies depending upon the sophistication in case of a permanent intake while in case of a temporary intake, it is related to construction and repair. Temporary intakes are generally washed away during floods. They have to be reconstructed after a flood to divert the water. The cost depends upon the type and size of the river, availability of construction material, frequency and intensity of floods in that season;
- *canals*. The problems in the canals in the hills and in the plains are different. Sliding, cross drainage and flood damage are the major maintenance problems in the hill canals while sedimentation and cross drainage are dominant problems in the plains;
- *command area*. Flood damage to the canal system and to the command area is also a major problem. The problem of flood comes basically from the sediment. Due to a high sediment load and unexpected deposition at some points the river changes its course and may damage an irrigation scheme;
- *desilting basin*. Removal of sediment from a desilting basin is one of the major O&M activities. Sometimes this becomes the bottleneck for the scheme as overall water availability relies entirely on the proper operation of the desilting basin.

The operation and maintenance cost can be greatly reduced by controlling the sediment entry into the scheme, controlled deposition at the appropriate locations or transporting the entered sediment as much as possible to the field. Hence understanding of the sediment transport problem in irrigation canals and design of non-silting and non-eroding canals is a vital issue for the modernization of schemes.

1.4   Sediment transport aspects in irrigation canals

Main objective of an irrigation engineer is to design an irrigation system that requires least or no maintenance. The canal should be capable of delivering the required amount of water to the targeted command area, without the need for excavation of sediment for the full life of the system. Dealing with sediment is one of the major difficulties in irrigation system design, yet one of the most important aspects.

Although some systems operate with essentially sediment free water, in many instances irrigation water is taken directly from a river in which there is a natural supply of sediment. By suitably introducing sediment excluding structures at the head of the canal, the coarser fraction is generally separated and returned to the river. The remaining sediment that enters into the canal would have to be transported to the fields. The sediment transporting capacity of the canal for changing flow conditions must match with the supply of sediment. Transporting the whole range of quantities and size of sediment requires rapid flows with relatively steep slopes, which is generally not possible in unlined earthen canals. Also in most cases the topography restricts providing a relatively steep slope. Hence, it is necessary to know in advance the size and quantity of sediment the canal is capable of transporting so that a suitable sediment removal facility can be installed at the head of the canal.

1.4.1   Canal design

The design of irrigation canals for sediment laden water would have to include aspects related to the irrigation criteria as well as to the sediment transport. The need of conveying different amounts of water to meet the irrigation requirements for a required water level is the main criterion for canal design (Dahmen, 1994). Furthermore, the design must be compatible with a particular local sediment load in order to avoid silting and/or scouring. The water supply would have to meet the irrigation requirements and at the same time the least deposition and/or erosion should occur in the canals.

Therefore the design of stable irrigation canals needs to consider the full operating life of the irrigation network and would have to be based upon the requirement that the total sediment inflow during a certain period is equal to the total sediment outflow. Sediments may be deposited during one phase of the irrigation season and be eroded during another phase, but there needs to be an overall balance of erosion and deposition for the whole operation period.

The design of a canal with a certain sediment load requires a set of equations related to the water-sediment flow to provide the unknown design variables of bed slope and cross section (bed width, water depth and side slope). The geometry of a sediment carrying irrigation canal will be the end product of a design process in which the flow of water and sediment transport interacts.

Modernization of irrigation schemes

Modernization is sometimes mixed with rehabilitation. There is a clear difference between rehabilitation and modernization. Rehabilitation works are carried out to meet the original objectives for which the scheme was built, while modernization refers to improvements in the scheme to meet the new objectives and new type of services (Easter, *et al.*, 1998).

The design of an irrigation scheme for sediment transport needs different approaches depending upon whether the scheme is a new one or a to be modernized one. In case of a new scheme it might be possible to design the scheme for efficient sediment management. Modernization of existing schemes poses some extra complications that make it more difficult for the design. Major difficulties in the modernization type of schemes are:

- *slope*. Existing alignment and levels and position of the off-takes provide very little flexibility of adjusting the slope to make it optimal for sediment transport;
- *structures*. Some structures are efficient in conveying sediment while others are not. The structures should have minimum disturbance to the flow pattern, so that the sediment carrying capacity of the canal is not reduced. Schemes that are designed and run on the basis of imposed water allocation and proportional delivery schedules require few flow control structures. To ensure flexible, efficient and reliable distribution there would be an obvious increase in the number of control structures. Provision of more control structures and frequent and time dependent operation of these structures to meet the irrigation demand makes the flow highly non-uniform. Hence the sediment carrying capacity of the canal is changed significantly. Selection and design of suitable flow control structures that meet the water delivery schedule and at the same time cause least sediment deposition is the major difficulty;
- *water delivery schedule*. The water delivery schedule will also have a significant effect on the sediment transport behaviour of the canal. It basically dictates the operation of flow control structures, which ultimately affect the sediment transport. The water delivery schedule is decided by the cropping pattern, capacity of the canal, available water and water rights. However, during modernization there is a possibility to include the sediment transport criteria also in the determination of the water delivery schedule;
- *management of the scheme*. It may not be possible to design all the canals with equal efficiency of sediment transport. Some social and management problems may come up if the scheme has to be managed jointly by the agency and the Water Users Association and especially when there are a number of small groups responsible for the management of separate canals.

Modernisation of a scheme to meet the crop water requirement in a more flexible way would help greatly in making the scheme sustainable. Design of efficient canal systems is a major part of this activity. It would be a timely and right decision to find out the methods of dealing with this aspect to reduce the risk of going through the same cycle of rehabilitation, deterioration and again rehabilitation in the future.

1.4.2   Operation and management

Irrigation schemes that carry a sediment load are not easy to operate without compromising in flexibility or maintenance cost (Horst, 1998). The assumptions made in the design are mostly very difficult to achieve in the real life. Any deviation from the assumed conditions results in the inability of the scheme to run properly. Hence it is very important with respect to sustainability, to know beforehand, what would be the effect if the assumptions are not met.

One of the quick and the cheapest way of simulating such scenarios is by mathematical modelling. In-depth understanding of the water and sediment flow process and the effect of changing flow, sediment concentration, maintenance conditions and operation of flow control structures to meet the water delivery schedules on this process would be helpful in proposing a stable canal design.

1.5   Irrigation in Nepal

Irrigation is, in many ways, a major factor in the development of Nepal. It is the largest water use sub-sector, affects the life of many people involved in agriculture, is the major contributor (40%) to the Gross Domestic Product (GDP) (World Bank, 2005) and a major factor for maintaining food security in the country. The Government, from the very beginning, has wisely recognized this fact and given due importance in its yearly and five year plans. In the 10th five-year plan (2002-2007), 9.7% of the total national development budget was allocated to irrigation. Given the importance of irrigation and large investments already made and planned for the future, the effectiveness of water delivery and its ultimate sustainability are of major concern. The Water and Energy Commission Secretariat indicates that many schemes have not reached their planned level of productivity and are not sustainable, financially as well as technically (Water and Energy Commission Secretariat, 2003).

In its plan for the future, the Government wants the increment in irrigated area by constructing new schemes and at the same time it is concerned about the efficiency, coverage area, cropping intensity as well as recovery of operation and maintenance costs of the existing irrigation schemes. Two types of activities would continue side by side for irrigation development, firstly modernization of the already constructed irrigation schemes and secondly construction of new schemes to provide irrigation to more land. The focus now is on modernization and the objectives of these works are:
– increase the performance of the schemes by providing more efficient water delivery;
– reduce the operation and maintenance costs by addressing the sediment control and transport issues;
– organise the farmers and impart training for scheme operation and maintenance as well as on farm water management;
– provide agriculture extension services to increase the productivity;
– transfer the management of the scheme to the farmers depending upon their capability.

Most of the schemes in Nepal are supply oriented whose objectives were to distribute irrigation water to the maximum number of farmers. The design capacity of the canals is low (for example the duty of the Sunsari Morang Irrigation Scheme is 0.67 l/s.ha) for intensive irrigation. The trend is to overestimate the available water in the source and include more area under command due to the social pressure. The problem is augmented by seepage losses and reduced canal capacity due to sedimentation.

Most irrigation schemes are run-of-river type carrying a high sediment load. Due to insufficient maintenance theses schemes are highly unreliable and inadequate in terms of water delivery. A performance study of five Government built and operated irrigation schemes showed no difference in cropping intensity and yield as compared to the non-irrigated areas (Water and Energy Commission Secretariat, 1982). During modernization the previous water distribution provisions of un-gated overflow weirs have generally been replaced with gates and converted into orifices. The objectives of such changes were to increase flexibility in water delivery and to increase equity. After the introduction of a large number of manually operated gates, the operation and maintenance cost increased sharply and could not be met by the annual budget (Pradhan, 1996). Further, operation of gates changed the sediment transport pattern leading to the unexpected erosion/deposition along the canal network. Sedimentation increased maintenance cost and in the absence of proper maintenance the targeted service level could not be achieved (Khanal, 2003).

To achieve the target set by the Government for modernized as well as new irrigation schemes an improvement in the design concept that is compatible with the proposed management mode is necessary. Since sediment has been the major concern for the sustainability of irrigation schemes, this aspect needs to be properly included in the formulation of management modes and accordingly the schemes need to be designed.

1.6   Mathematical modelling

Modelling of irrigation canals for sediment transport involves the solution of water flow and sediment transport equations. The computational environment for canal simulation is much more demanding than for river flow due to the extreme variability and unsteadiness of the flow, the presence of numerous hydraulic structures, dynamic gate movements and pump operations, and possible topographical complexity. The flow in irrigation canals is often entirely in the sub-critical region, although supercritical flow may occur in some parts of the canal system. A supercritical flow is accompanied by a downstream hydraulic jump. A hydraulic jump may be stationary when a physical structure is present to stabilise the jump location. In that case a depth discharge relation of the structure can be used as boundary condition for the flow calculations. It is fairly logical in irrigation canals to assume that every supercritical flow is accompanied by a structure. Irrigation canals have specific properties that can be used to simplify the equations defining flow and sediment transport.

The study of operational plans, maintenance conditions on the sediment transport behaviour of the canal and transfer of this knowledge to the farmers and managers in terms of efficient water management is vital for the sustainability of the scheme. As this aspect is directly related to the reduction of O&M cost and equity, adequacy and reliability a better understanding of sediment transport processes under different operation, management and maintenance conditions for the designer, managers and trainers is necessary. Until now in practise sediment transport generally has been considered to be beyond normal understanding and control. With the introduction of mathematical modelling this aspect can be better explained. However, the reliability of predictors is still poor, but even then the simulation can show the difference of one management plan over the others. Hence investigation and formulation of a design tool for irrigation canals might be a necessary and timely step.

1.7   Objectives

Main objective of the research was to understand the relevant aspects of sediment transport of irrigation canals and to formulate a design approach that includes sediment management. In line with the main objective the following underlying objectives were set:
– to understand the different design methods used in the design of irrigation schemes in Nepal and their effectiveness in sediment transport;
– to evaluate the effect of water delivery schedules, flow control systems and operation and maintenance plans on the sediment transport capacity of the irrigation canals;
– to suggest a canal design procedure that is based on the holistic concept of changing discharge and sediment inflow;
– to verify the procedure with field measurement data from an irrigation scheme in Nepal.

To achieve the aforesaid objectives the following specific activities have been carried out:
– analysis of the velocity and shear stress distribution across the trapezoidal canal and its effect in the sediment transport;
– investigation of the methods of computing equivalent roughness in trapezoidal canals and the process of including the effects of vegetation and maintenance activities in the roughness determination process;
– improvement and testing of the mathematical model SETRIC for the simulation of water and sediment flow in an upstream control type irrigation scheme including water management and the system maintenance;
– filed investigation and data collection regarding the design of canals, operation and maintenance and control of the irrigation scheme in view of sediment transport;
– use the field data to simulate and compare the existing design and management approach with the proposed design and management approach.

1.8   Set-up of the thesis

The thesis is expanded in 9 chapters. The first chapter provides the background information regarding the problem faced by the irrigation schemes in general. It also gives an account on the design and management aspects of irrigation scheme carrying sediment and the formulation of research objectives.

On the basis of the problem in formulating the design and management aspects of irrigation schemes that carry sediment, the objectives of the research were formulated.

Chapter 2 provides an overview of history of irrigation, status, issues and future plans of irrigation development in Nepal. A brief introduction of the Sunsari Morang Irrigation Scheme that is selected for data collection is also given.

Chapter 3 gives the review of canal design methods, water flow hydraulics, sediment transport aspect under equilibrium and non-equilibrium conditions, modelling aspects of sediment transport and an overview of available sediment transport models.

Chapter 4 gives an assessment of derivation of design discharge, estimation of roughness, sensitivity of selected sediment transport predictors with the change in canal geometry, the velocity and shear stress distribution in an irrigation canal, the effect of changing velocity and shear in the sediment transport process and the proposal for applying correction to the selected sediment transport predictors.

Chapter 5 focuses on the evaluation of existing canal design approaches in Nepal, their limitations, the rationale and the steps of the proposed design approach and the management aspect of the canal design.

Chapter 6 gives a description of the mathematical sediment transport model SETRIC, theoretical background, model formulation concept, schematization of various processes, computations steps and limitations.

Chapter 7 describes the water management plans and the design aspects of Secondary Canal S9, morphological, water flow, scheme management and sediment data collection, analysis of the scheme operation plan and the evaluation of the collected data.

Chapter 8 is about the application of mathematical model to evaluate the proposed design approach and comparison results with the existing canal.

Chapter 9 gives the evaluation of conclusions drawn from the inferences of previous chapters and some outlook for the future in this field.

2
Irrigation in Nepal

2.1   General overview

Nepal is a landlocked country in South Asia lying between China and India. It is situated between 26°22' N to 30°27' N latitude and 80°4' E to 88°12' E longitude of the prime meridian. Roughly rectangular in shape, the country has an area of 147,181 km^2. It is 885 km in length but its width is uneven and increases towards the West. The mean North-South width is 193 km. Nepal is a predominantly mountainous country, with elevations ranging from 64 m+MSL (Mean Sea Level) at Kechana in the eastern Terai District of Jhapa to 8,848 m+MSL at the peak of the world highest mountain, Sagarmatha[1], within a span of 200 km.

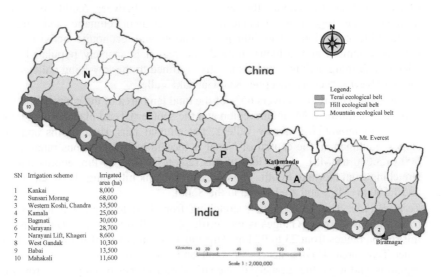

SN	Irrigation scheme	Irrigated area (ha)
1	Kankai	8,000
2	Sunsari Morang	68,000
3	Western Koshi, Chandra	35,500
4	Kamala	25,000
5	Bagmati	30,000
6	Narayani	28,700
7	Narayani Lift, Khageri	8,600
8	West Gandak	10,300
9	Babai	13,500
10	Mahakali	11,600

Figure 2.1   Map of Nepal showing ecological zones and major surface irrigation schemes (Department of Irrigation, 1990b, Poudel, 2003).

The climate varies in a wide range due to the diverse topography. Summer and late spring maximum temperatures range from more than 40 °C in the Terai to about 28 °C in the central part of the country, with May being the warmest month. Winter average maximum and minimum temperatures in the Terai range from 23 °C to 7 °C, while the central valleys experience a 12 °C maximum and a below freezing minimum. Much colder temperatures are experienced at higher elevations.

Rainfall in the country is primarily due to southwest monsoon during summer from June to September that accounts for about 80% of the annual precipitation and the remaining 20% during the remaining 8 months. In the hydrological cycle, about 64% of all rainfall immediately drains as surface runoff. Of the 36% some is

---

[1] Sagarmatha is the Nepalese name of Mt. Everest.

retained in the form of snow and ice in the high mountains, some percolates through the ground as groundwater and some results in evaporation and transpiration (Water and Energy Commission Secretariat, 2003).

Mean annual precipitation ranges from more than 6,000 mm along the southern slope of the Annapurna mountain range in the central part of the country to less than 250 mm in the northern central portion near the Tibetan Plateau. Most of the country experiences annual precipitation in the range of 1,500 mm to 2,500 mm, with a distinct maximum along the southern slopes of the Mahabharat[2] and the Himalayan ranges[3] in the eastern two third of the country. Based on available hydrological data, the estimated annual runoff from the rivers of Nepal is 220 billion m^3, with an average annual precipitation of 1,530 mm. According to its topography Nepal can be roughly divided into three parts:

– *Mountain Region.* This occupies roughly 15% of the country's total area. Stretching from East to West its width ranges from 24 to 28 km. It is situated at 4,000 or more m+MSL to the north of the Hill Region. It constitutes the central portion of the Himalayan range originating in the Pamirs, a high altitude region of Central Asia. In general, the snow line occurs between 5,000 and 5,500 m+MSL. The region is characterized by inclement climatic and rugged topographic conditions. The human habitation and economic activities are extremely limited and arduous. Indeed, the region is sparsely populated, and whatever farming activity exists is mostly confined to the low-lying valleys and the river basins, such as the upper Kali Gandaki Valley;

– *Hill Region.* This region covers 65% of the total land area of the country and lies between the northern Mountain region and the southern Terai plains. This region consists of mountains, narrow valleys, flat alluvial fans, river basins and hill slopes. Kathmandu, the capital of Nepal is located here. Elevations range from 500 to 3,000 m+MSL. The width of this region is 64 to 84 km, approximately stretching from North to South. The Chure or Shivalik hills range between 700 m and 1,500 m+MSL and are located in the South;

– *Terai Region.* The southern belt stretching from East to West and formed by tributaries of the river Ganges is the fertile plain called the Terai. The width of this region ranges from 23 to 97 km. This range is composed of flat lands and the inner Terai area. Having access to water and endowed with a fertile soil, this region is said to be the 'granary of Nepal'. Various types of crops are grown in this area. Since the plain terrain has facilitated the exploitation of natural resources transportation, communication and energy production have reached a comparatively developed form in this region.

2.2   Land and water resources

Approximately 6,000 rivers and rivulets, with a total drainage basin of 194,471 km^2 flow through Nepal; 76% of this drainage basin is contained within Nepal. Drainage basins of 33 of Nepal's rivers are greater then 1,000 km^2. All the rivers from Nepal are the tributaries of the Ganges, one of the twenty largest rivers in the

---

[2] Mahabharat is the mountain range that lies in the central part of the country stretching from East to West. Elevations of the highest peaks in this range are in the order of 1,800 m AMSL.
[3] Himalayan ranges are snow caped mountain ranges in Asia, ranging from Pakistan, India, Nepal, Bhutan to Sikkim of India.

world. These tributaries contribute 40% of the mean annual flow and 70% of the dry season flow of the Ganges (Water and Energy Commission Secretariat, 2003).

There are three types of rivers in Nepal, classified on the nature of their source and discharge. In the first category are perennial rivers that originate in the Himalayas and carry snow fed flows with a significant discharge, even in the dry season. These include the Koshi, Gandaki, Karnali and Mahakali river systems. In the second category are the Mechi, Kankai, Kamala, Bagmati, West Rapti and Babai rivers, which originate from the Midlands or Mahabharat range of mountains and are fed by precipitation as well as groundwater regeneration, including springs. Although these rivers are also perennial, they are commonly characterized by a wide fluctuation of discharge. The third category includes a large number of small rivers in the Terai (southern plains), which originate from the southern Siwalik range of hills. These rivers are seasonal with little flow during the dry season, but characterized by flash floods during the monsoon.

2.3   History of irrigation development

The history of irrigation development in Nepal in view of issues related to construction and water management is summarized in this section. Historians divide the history of Nepal into three periods: Ancient Period from 500 B.C. to 700, the Medieval Period from 750 to 1750, and the Modern Period.

2.3.1   Ancient Period

The history of using river water for agricultural purposes is very old in Nepal. Security from famine and hunger might have necessitated the construction of irrigation schemes and accordingly indigenously built and managed irrigation schemes are in existence in Nepal. Buddha (500-600 BC) played a mediating role in the settlement of a water distribution dispute of Rohini river between the Shakya and Kolya communities around 2500 years ago (Pesala, 2006), however traces of the irrigation infrastructure of that era have so far not been found. With the introduction of brick manufacturing techniques in the 5th century permanent types of irrigation infrastructure were constructed. Infrastructure constructed after that period could be seen in Kathmandu valley (Poudel, 2003). The rulers took the initiative in the construction of irrigation canals as it not only helped to produce more food and created security against famine but also was a source of revenue (Agrawal, 1980). Those irrigation schemes constructed with the compulsory labour contribution did not receive Government assistance for their construction and maintenance. Rules and regulations made for their operation and maintenance became part of culture and tradition (Regmi, 1969).

2.3.2   Medieval Period

The Medieval Period is considered from 750 to 1750. The technology and tradition of construction of irrigation and irrigated farming slowly transferred from Kathmandu Valley to the surrounding areas and it is believed that irrigated agriculture, especially paddy cropping, started in the mid and high hills of Nepal by the beginning of 1000. The importance of cultivable and fertile land increased slowly. The concept that the irrigation and its management comes under the jurisdiction of farmers organizations and that the Government should not intervene into it was in place as early as 1674. They also motivated farmers/communities to take a lead role in dispute management and mediation (Poudel, 2003). Argheli Kulo (canal) of Western Nepal was constructed in this period and is still functioning well.

The strength of a state used to be decided by the area of cultivable land and the boundary of land used to be dictated by military activity. Un-irrigated sloping lands in the hills were cultivated after potato and maize were introduced by the British in this region. The 18th century could be considered as the golden age for community and farmers constructed and managed irrigation schemes. To some extent the Government also provided some support to such irrigation schemes. These schemes are popularly known as Farmers Managed Irrigation Schemes (FMIS) and are functioning well till now (Khanal, 2003).

2.3.3   Modern Period

The Modern Period can be divided into the pre-planning period (before 1957) and the planned development period (after 1957). In 1864, the 1st Civil Code (Muluki Ain) was promulgated, which codified the age old customs and cultures of different casts and communities related to water management and helped to institutionalize the hydraulic traditions of Nepal (Höfer, 1979). This law was a milestone in developing institutionally sound, self sustainable irrigation systems that have been the examples of community managed irrigation schemes in the world. This legal document reflected the diverse requirements of different places and its linkage to the culture and religion of several communities in terms of irrigation management. This also helped in recognizing the cultural diversity, special requirements at local level and linkage of different communities to settle the dispute of water allocation that has established a guideline to formulate the rules and regulations of modern irrigation development. During that period Nepal contributed 100,000 ha to the 8,000,000 ha of total irrigated land of the world (Poudel, 2003).

By the beginning of 20th century, the Government started to construct and manage the irrigation schemes, thus ending the age old tradition of non-involvement in irrigation construction and management affairs. Construction of irrigation schemes with Government initiation and investment started with the construction of the Chandra canal in 1922. This irrigation scheme with a command area of 10,000 ha was designed and constructed by British engineers in India (Department of Irrigation, 2000). Until 1954 no major activities were undertaken for irrigation development as in this period only two schemes with a total command area of 3,000 ha were constructed. The two schemes were Jagdishpur Canal (renamed later as Banaganga Irrigation Scheme) with a command area of 1,000 ha and Judha Canal (renamed later as Manusmara Irrigation Scheme) with a command area of 2,000 ha.

After 1950 the tendency of intervention into the community managed irrigation schemes by a Government agency in terms of investment in construction as well as in operation and maintenance increased gradually. More and more isolated small irrigation schemes were integrated into large-scale schemes and Government officials played a key role in determining the construction, maintenance and operation plan of the scheme. In view of the increased irrigation activity the

Irrigation Department was established in May 1952 under the Construction and Communication Ministry (Parajuli and Sharma, 2003).

In eastern Terai, the Government contributed fifty percent of the total expenditure while the remaining fifty percent had to be contributed by the farmers. The army was given the responsibility of operation and maintenance work under the coordination of Government officials. Small maintenance works used to be done by the farmers while the work out of the capacity of farmers was done by the army. Local leaders were given the responsibility of operation, maintenance and revenue collection. Each area had a fixed revenue and the responsible local leaders had to add from their side the remaining amount in case they could not collect the targeted amount. This prompted them to upkeep the irrigation scheme and pay special attention in the maintenance and strictly follow the rules of water distribution.

Major construction of large-scale irrigation schemes in the Terai started with the bilateral agreement with India (Koshi agreement of 1954 and Gandak agreement of 1959). As a result of these agreements Chatra (later renamed as Sunsari Morang Irrigation Scheme) (68,000 ha), West Koshi (21,000 ha), Narayani (38,000 ha) and West Gandak (13,000 ha) were designed and constructed (Department of Irrigation, 2007). Though, the Government had recognized the importance of irrigation for increasing agricultural production at this stage the country did not have the adequate technical manpower and financial resources to implement large-scale irrigation works. So in the first five year plan (1957-1962) the priority was to construct small to medium scale irrigation schemes, especially in the hills with the focus on the participation of the users.

Table 2.1   Summary of expenditure and increment in irrigated area (Hada, 2003, National Planning Commission, 2007).

Development plans	Irrigated area (ha)			% of total irrigable area*	Expenditure (millions US$)
	Government	Farmers	Total		
1. 1956-1961	5,200		11,428	0.6	2
2. 1962-1965	1,035		12,463	0.7	7
3. 1965-1970	52,860		65,323	3.7	13
4. 1970-1975	37,733		103,056	5.8	26
5. 1975-1980	95,425		198,481	11.2	72
6. 1980-1985	140,191		338,672	19.2	219
7. 1985-1990	128,730		467,402	26.5	201
8. 1992-1997	206,401	381,814	1,055,617	59.8	280
9. 1997-2002	146,703	300,935	1,121,441	63.5	449
10. 2002-2007	241,000	236,935	1,298,441	73.5	257

* Total irrigable area 1,766,000 ha

The Government investment in irrigation development - especially in the large-scale irrigation schemes in the Terai increased tremendously from 1970 onwards. This was due to the increase in the borrowing of international capital in the form of loans and grants for the country's overall economic development. This is clearly reflected in the surge of irrigation development targets in the subsequent five-year development plans from the Fourth Plan (1970-1975) onwards. The main focus of the irrigation development up to 1980 was the construction of the physical infrastructure. While after 1985 the focus shifted from constructing new irrigation schemes to improving management and modernization of the schemes to achieve the targeted objectives.

Table 2.1 gives the summary of irrigation activity by the Government during the last ten development plans. From the 8th five year plan (1992-1997) rehabilitation works of Farmers Managed Irrigation Schemes (FMIS) started and during this period around 80,900 ha area was rehabilitated, while only 65,900 ha rainfed area was brought under irrigation. In the 10th plan (2002-2007) the plan was to rehabilitate 64,000 ha.

Nepal has a cultivated area of 2,642,000 ha (18% of its land area), of which two third (1,766,000 ha) is potentially irrigable. At present 42% of the cultivated area has irrigation of some sort, but only 17% of cultivated area has year round irrigation[4] (i.e., only 41% of the irrigated area gets year round irrigation). In the Terai, 82% of the total irrigated area (889,000 ha) is through surface irrigation and the remaining 18% is through groundwater. Most of the irrigated areas (and the future potential) are situated in the Terai, the fertile lowland. It is estimated that existing irrigation schemes contribute approximately 65% of the country's current agriculture production (Water and Energy Commission Secretariat, 2003) as compared to the 40% crop output from 18% irrigated land in the world (Schultz, 2002).

Table 2.2   Irrigation development in Nepal 1999/2000 (Water and Energy Commission Secretariat, 2003, Central Bureau of Statistics, 2006).

Geographic region	Total area (10^3 ha)				Irrigated as % of cultivated	Year round irrigated as % of irrigated
	Cultivated	Irrigable	Irrigated	Year round irrigation		
Terai	1,360	1,338	889	368	65	41
Hills	1,054	369	167	66	16	39
Mountains	227	60	48	18	21	38
Total	2,642	1,766	1,104	452	42	41

As shown in Table 2.2 around 62.5% of the potentially irrigable area has been provided with irrigation facilities. To irrigate the remaining area large-scale multipurpose land and water development projects have to be implemented that need huge investments, otherwise they are economically less viable. Even the developed existing schemes are running far below their target level. Around 90% of the command area is covered during the wet season while the coverage is only around 25% during the dry season. This is in fact due to the high seasonal variation of available water in the streams. Most of small and medium scale schemes take water from small to medium sized streams and are affected most. The coverage during the dry season is nominal even for large-scale schemes from snow-fed perennial rivers due to design and operational limitations. Even if the large-scale multipurpose

---

[4] As per Water and Energy Commission Secretariat (WECS) an irrigation scheme that has 150% or more cropping intensity is said to have year round irrigation.

storage and inter-basin river diversion projects were implemented with great urgency over the next twenty-five years, there would still be 55% of the irrigable arable land non-irrigated between November and May in 2025 (Shah and Singh, 2001).

There is an urgent need for the optimal use of the available water in order to increase the agricultural production. The productivity of the irrigated area in Nepal is one of the lowest in South Asia. There are a number of factors behind this low level of production but reliable irrigation is the key issue. Reliability is a relative term and a scheme supposed to be reliable twenty years ago may not be rated as good now. To meet the demand of more food, the productivity has to be increased and at the same time to make the agriculture profitable high valued crops have to be grown. This demands for more flexible and assured water that is difficult with the present level of the infrastructure and the design and operation concepts.

2.4   Issues of the irrigation sector

The Water and Energy Commission Secretariat (2003) through a series of interaction workshops between experts, irrigation professionals, planners and political leaders have identified some key issues in the irrigation sector that can be summarised as:
- reorientation of supply driven approach;
- poor performance of irrigation schemes;
- lack of effective implementation of the Agricultural Perspective Plan (APP)[5];
- farmer's dependency syndromes and sustainability;
- problems of river management;
- weak institutional capability;
- symbiotic relationship between agriculture and irrigation (weak linkages);
- strengthening of Water Users Associations (WUA).

Short-term and long-term land and water development strategies have been prepared accordingly to address the issues identified. In the 10th five year plan (2002-2007), targeted development strategy for irrigation included:
- year round irrigation to 50% of irrigated land;
- 40% increase in average cereal yield in irrigated areas;
- establishment of WUAs that are capable of managing irrigation schemes up to 500 ha and all Agency Managed Irrigation Schemes (AMIS), jointly managed with WUAs;
- an average cropping intensity that exceeds 200% in year round irrigated areas.

---

[5] Agricultural Perspective Plan (APP) is a 20-year framework for agricultural development prepared by His Majesty's Government (HMG) with assistance from the International Development Agency (IDA). It was launched by HMG in 1997 and it is expected that, as a result of APP implementation, approximately 612,000 ha of irrigable land will eventually be converted to year round irrigation by emphasizing tube-well irrigation in the Terai. By the end of the APP in 2016/2017, it is expected that irrigation will command about 1.44 million ha or 55% of the cultivable land in Nepal.

In the next ten years (by 2017), the targeted achievements include:
- establishment of WUAs that are capable of managing irrigation schemes up to 5,000 ha;
- provision of year round irrigation to 66% of the irrigated areas;
- provision of irrigation schemes to 80% of all irrigable land;
- 125% increase in average cereal yields in irrigated area;
- an increase in the effective use of command area to 80%;
- an Irrigation Service Contribution (ISC) by farmers that exceeds 20%.

Targets to be achieved by the end of the 25 years (by end of 2027) strategy period include:
- provision of irrigation services to 90% of irrigable lands;
- an average cropping intensity that exceeds 250%;
- an increase in irrigation scheme efficiency to 60%;
- an increase in the effective use of command areas to 100%.

To carry out the strategy and achieve these targets, actions will be taken to:
- integrate irrigation planning and management with agricultural development;
- improve management of existing irrigation schemes;
- improve planning and implementation of new irrigation schemes;
- develop year round irrigation in support of intensification and diversification of agriculture;
- strengthen local capacity for planning, implementation and management of irrigation;
- encourage consolidation of land to promote irrigation/agriculture efficiency;
- improve groundwater development and management.

2.5   Description of the study area

2.5.1   Location and development

The Sunsari Morang Irrigation Scheme (SMIS) is located, some 240 km southeast of Kathmandu, lying along the international boundary with India, in Sunsari and Morang districts of the Koshi Zone in the Eastern Development Region of Nepal. The Koshi River is the source of water for the scheme (Figure 2.2). The drainage basin of this river lies between the central part of Nepal in the West and Indo-Nepal border in the East. The Arun and Sunkoshi, two of the major seven tributaries of Koshi River, originate from Tibet. The Koshi River has a drainage basin of 59,550 km^2 above the intake location of the scheme. After that point it emerges into the plain and flows for a length of 307 km (53 km in Nepal) before it joins the river Ganges near Kursela in Bihar, India. The peak discharge observed in the river was 25,900 m^3/s on 5th October 1968. The discharge goes as low as 240 m^3/s during dry periods.

In 1954, Nepal and India agreed to finance and construct the Chatra Canal Project (CCP) in Nepal. The Chatra Canal Project's objective was to irrigate 68,000 ha (net) land of Sunsari and Morang districts in the eastern Terai. The required water had to be diverted from the left bank of the Koshi River through a side intake. The construction of the scheme started in 1964 and after trial operation for two years, it was handed over to Nepal in 1975 (Department of Irrigation, 1995b). The Chatra Canal Project was later renamed as Sunsari Morang Irrigation Scheme. General features of the scheme (shown in Figure 2.3) as constructed and handed over to Nepal were:

- *canal system*. The main canal was designed for a discharge of 45.3 m^3/s. The total length of the canal was 52.7 km. It was unlined in the whole length. There were 10 secondary canals, 9 minors (off-take from the main canal that have a command area of less than 1,000 ha) and large numbers of direct outlets for irrigating the targeted command area. The canal slope was flatter than that required from Lacey's regime formula (Department of Irrigation, 1995a), so during the design, either the sediment transport criteria was not considered or a smaller silt factor were selected;
- *diversion requirement and duty*. The design duty of the canal was 0.67 l/s.ha, which was insufficient to irrigate the command area. With the 80% probable rainfall only around 50% of the targeted command area could be provided with secured irrigation. A side intake was constructed to supply the required 45.3 m^3/s;
- *control structures*. Very limited numbers of water level regulators were provided in the main canal. The off-takes (secondary and minors) were provided with gated undershot type of discharge regulators. No discharge measurement structures had been provided in the main canal. Regulators were calibrated for this purpose that did not furnish accurate values due to deposition of sediment upstream and downstream of the structures.

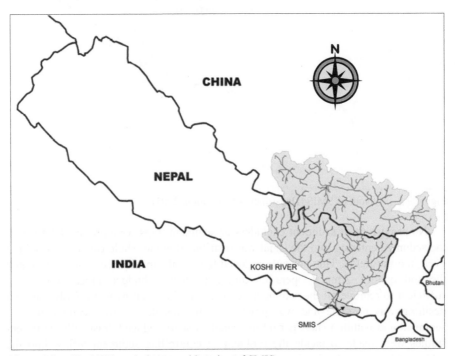

Figure 2.2   Koshi River drainage and location of SMIS.

2.5.2   Major problems

The irrigation scheme as constructed by the Government of India terminated at turnouts each serving an area of over 200 ha without a minor canal system below this level. The scheme therefore could not properly irrigate the whole area. The figure of 0.67 l/s.ha represents the design philosophy of that time as indicated earlier when coverage and protection from crop failure was more important than the crop water requirement criteria.

During the wet season (July to October) a lot of sediment used to enter the canal network and settled in the main and secondary canals, reducing the capacity of the scheme drastically. While during the dry season, the water level in the river went down and it proved to be difficult to divert the required amount of water. Moreover the river itself was shifting away from the intake, which made it even more difficult to divert the required amount of water.

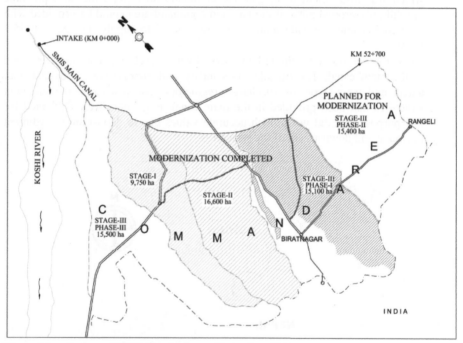

Figure 2.3   Layout of SMIS (Department of Irrigation, 2003).

No provisions of sediment excluders or sediment traps were provided; hence all the sediment entering into the canal was distributed in the whole canal network, and even into the fields. This not only reduced the canal capacity, but also deteriorated the soil quality. Annual operation and maintenance budget allocated was not sufficient to remove all the sediment deposited in the canal network. Only a partial desilting was possible and was practiced. This resulted in the accumulation of sediment and within a short period the canal capacity reduced drastically. Different studies were made to tackle the problem of controlling sediment inflow into the scheme (Department of Irrigation, 1982, 1985a, 1985b) as well as to reduce the sediment deposition in the canal network (Department of Irrigation, 1979, HR Wallingford, 1988).

2.5.3   Modernization works

Due to the above-mentioned reasons, the scheme as originally built did not fulfil its objectives, and the Government requested for the International Development Agency's assistance to overcome the shortcomings of the Sunsari Morang Irrigation Scheme as early as 1978, only three years after the hand-over (1975). The International Development Agency approved a credit of US$ 30 million to rehabilitate the scheme and to carry out a command area development of 9,750 ha. This development program for the 1st stage, known as SMIP-I, was completed in 1987 (Department of Irrigation, 1995a). In this phase mostly three types of works were carried out to address the problems:
–   construction of sub-secondary and tertiary canals up to a block area (tertiary unit) of 50 ha in 9,750 ha;
–   removal of sediment from the canal system;
–   installation of vortex tubes for sediment extraction.

In 1987 the International Development Agency approved a second stage of the project (known as SMIP-II) for the amount of US$ 40 million for the command area development (16,700 ha) and to deal with the sediment problem. At this stage the block area was reduced to 28 ha for the development of the tertiary canal system (Department of Irrigation, 1987). In addition to that main and secondary canals were rehabilitated.

The steps taken to reduce the sediment problem were not effective due to the operational problems as a lot of the already scarce water had to be drained to flush the sediment. Moreover, the diversion of water towards the canal became increasingly difficult due to the shifting of the river course away from the intake. It was realised that unless the water diversion and sediment problem would be dealt with properly, the command area development only will not yield substantial benefits.

In 1991, the Government obtained the advice of an International Panel of Experts (IPOE). IDA and the Government agreed with the IPOE proposals (Department of Irrigation, 1998), which included:
–   move the intake structure 1,300 m upstream;
–   construct a desilting basin;
–   utilise dredgers for removal of silt from the basin;
–   construct a micro-hydro unit in the main canal to provide power for the dredgers operation.

A supplementary IDA credit of US$ 28 million was approved in 1993 under the Sunsari Morang Headwork Project (SMHP). The work started in 1993 and was completed in May 1996 (Department of Irrigation, 1998). All the works were primarily focused to deal with the sediment problem and to assure sufficient water. The intake capacity has also been increased from 45.3 m^3/s to 60 m^3/s. The new intake structure has a provision of taking water from the top layer when the water level in the river increases to reduce the sediment entry into the system. After the implementation of this measure the sediment problem in the scheme has been reduced (Figure 2.4).

Figure 2.4   Relocated side intake (left) and the settling basin with continuous sediment removal arrangements (right).

In 1997 a third program started for a total cost of US$ 30 million (World Bank, 1997) and was completed in 2002. Major works of the proposed package included a command area development of 15,200 ha, flood protection works in the Koshi River and strengthening of the main canal. The details of the works completed in first phase of the third program are (Department of Irrigation, 1995a, 2003):
– command area development of 3,300 ha under the Biratnagar Branch Canal;
– command area development of 11,900 ha under the Harinagara Mahadeokaula Branch Canal;
– re-construction of the Budhikulo Aqueduct cum Bridge;
– strengthening of the Chatra Main Canal;
– construction of 5.5 km of the Koshi Flood Embankment on the left bank of Koshi River.

As more water was available after improvement of the headwork it was proposed to develop and extend 46,000 ha thus providing irrigation facilities to a total area of 73,000 ha. The study showed that with the discharge of 60 m^3/s the proposed command area could be irrigated with 70% probability (Department of Irrigation, 2003). However, the remaining two phases that cover around 28,000 ha are still to be implemented.

3
Design of canals for sediment transport

3.1 Background

Canals are used for the conveyance of water. A network of irrigation canals is needed to carry water from the source and to distribute it to the irrigated fields. A well planned, designed and constructed canal network is central in meeting the objectives of an irrigation scheme in a sustainable way. The design of a canal is a complex process of fixing its shape, slope and size based upon various aspects like amount and quality of water to be transported, type of canal to be constructed, the terrain through which it passes, socio-economic setting, climate, soil type, etc. The process becomes more complicated when the boundary of the canal is erodible and when the canal carries sediment with water. Unwanted erosion and deposition along the canal network has become a central problem for irrigation schemes and a lot of money and effort have been invested to find methods for stable canal design.

This chapter deals with the up to date studies and researches in the field of canal design and explores the hydraulic, sediment transport and modelling aspects of irrigation canals.

3.2 Design methods

From design perspective irrigation canals can be divided into three categories (Ranga Raju, 1981):
- *canals with rigid boundary*. The design of these canals is to ensure that the velocity is sufficient enough to transport any sediment entering into the canal. The velocity should not damage lining and create disturbances in the water surface;
- *canals with erodible boundary and carrying clear water*. The design of such canals involves finding the canal section such that the material in the boundary does not move;
- *canals with erodible boundary and carrying water with sediment*. The design should ensure that the velocity is high enough to convey all the sediment and at the same time not so high that the bed material is eroded. Hence, a balance between the transport of sediment entering into the canal and the stability of the boundary is maintained. This is the most difficult aspect of canal design.

Numerous researches have been done and a range of canal design methods have been developed and are in use around the world. In the field of alluvial canal design, i.e. the third category as per Raju (1981), the earliest work was of Kennedy (1895). Several others followed to find out the width predictors for alluvial canals. The concepts of minimum stream power (Chang, 1980), minimum energy dissipation (Brebner and Wilson, 1967, Yang, *et al.*, 1981) and maximum sediment transport (White, *et al.*, 1981b) have been proposed to overcome the limitations in the regime approach. Lane (1955) presented the tractive force theory developed by many others at the U.S. Bureau of Reclamation. This approach is more suitable for the second category of canals that carry very little sediment and the problem is limited to control the bed or bank erosion. Depending upon their fundamental solution these methods can be broadly classified under regime, tractive force and rational theories.

3.2.1   Regime theory

The regime design methods are sets of empirical equations based on observations of canals and rivers that have achieved dynamic stability. An alluvial canal is said to be "in regime" or dynamic stability when, over some suitably long period, its depth, width and slope stabilize to average or equilibrium values (Raudkivi, 1990). There may be seasonal deposition or erosion in the canal but the overall canal geometry in one water year remains unchanged. This can only occur when the sediment input to the canal matches the average sediment transporting capacity, sediment deposition during periods of high sediment input being balanced by periods of scour when the sediment input is low.

Regime methods were developed at the end of the 19th century to aid the design of major irrigation systems of the Indian sub-continent. The development started with Kennedy (1895) followed by Lindley (1919), Lacey (1930), Blench (1957), Simons and Albertson (1963). A critical analysis of these methods is given by Stevens and Nordin (1987) and they are well explained in the books by Shen (1976), Chang (1988), Raudkivi (1990) and Simons and Senturk (1992).

Among the regime methods one of the most popular set of equations is that of Lacey (1930). The equations were based on data from three canal systems of the Indian sub-continent. They specify the cross section and slope of regime canals from the incoming discharge and a representative bed material size. They are, with minor changes to coefficients and some redefinition of the silt factor, still widely used. The set of equations given by Lacey is:

$$f = 1.76 d_m^{\frac{1}{2}} \tag{3.1}$$

$$V_0 = 0.4382 \left(\frac{Q}{f^2} \right)^{(1/6)} \tag{3.2}$$

$$S_0 = 0.000315 \frac{f^{\frac{5}{3}}}{Q^{\frac{1}{6}}} \tag{3.3}$$

$$P = 4.83 Q^{\frac{1}{2}} \tag{3.4}$$

where

d_m	= mean sediment size (mm)
f	= silt factor
P	= wetted perimeter (m)
Q	= water flow rate (m³/s)
S_0	= bed slope (m/m)

V_0 = stable non silting velocity (m/s)

The limitation of this theory is that it assumes that the discharge is the only factor determining the wetted perimeter of the canal while the fact is that canals with less stable banks tend to be wider than those with strong banks for the same discharge.

Later on Simons and Albertson (1963) developed a set of regime equations based on a large data set collected from canal systems in India and North America. Five types of canals were identified from the collected data set and for each category the coefficients have been given for the design of a canal as shown in Table 3.1. Simons and Albertson's equations therefore have a wide range of applicability. The data set used related to a sediment load of less than 500 ppm (Raudkivi, 1990). The set of regime equations of Simons and Albertson (1963) is:

$$P = K_1 \sqrt{Q} \tag{3.5}$$

$$B = 0.9P = 0.92B_s - 0.61 \tag{3.6}$$

$$R = K_2 Q^{0.36} \tag{3.7}$$

$$h = 1.21R \qquad \text{for } R < 2.1 \text{ m} \tag{3.8}$$

$$h = 0.61 + 0.93R \qquad \text{for } R > 2.1 \text{ m} \tag{3.9}$$

$$V = K_3 \left(R^2 S \right)^n \tag{3.10}$$

$$\frac{C^2}{g} = \frac{V^2}{gDS} = K_4 \left(\frac{VB}{v} \right)^{0.37} \tag{3.11}$$

$$A = K_5 Q^{0.87} \tag{3.12}$$

where

A	= cross sectional area of the flow (m^2)	
B	= mean canal width (m) = A/h	
Bs	= water surface width (m)	
C	= Chézy coefficient (m$^{1/2}$/s)	
h	= mean flow depth (m)	
K$_i$	= coefficient (i = 1 to 5) for different canal types (ref Table 3.1)	
n	= exponent (ref Table 3.1)	
P	= wetted perimeter (m)	
Q	= flow rate (m^3/s)	
R	= hydraulic mean radius (m)	
S	= bed slope (m/m)	
V	= flow velocity (m/s)	
v	= kinematic viscosity (m^2/s)	

Chitale (1966) analysed data from the Indian sub-continent and fitted them to get equations similar to Lacey. He also studied the reason why many canals diverge from the Lacey dimensions and proposed Lacey divergence equations (Chitale, 1996). Ranga Raju *et al.* (1977) conducted laboratory flume experiments to determine the relative sensitivity of bed width and bed slope in self formed sand channels and concluded that the stable longitudinal slope was found to be extremely dependent on the sediment discharge. Stevens and Nordin (1987) gave an overview of the applicability and limitations of Lacey's regime theory and proposed a reformulation employing the laws of conservation of mass and Newton's law of motion.

Table 3.1   Value of coefficients in Simons and Albertson equations for different canal types (Simons, *et al.*, 1963).

Canal type	Coefficient					
	K_1	K_2	K_3	K_4	K_5	n
1. Sand bed and banks	6.34	0.4-0.6	9.33	0.33	2.6	0.33
2. Sand bed and cohesive banks	4.71	0.48	10.80	0.53	2.25	0.33
3. Cohesive bed and banks	4.0-4.7	0.41-0.56	-	0.88	2.25	-
4. Coarse non-cohesive material	3.2-3.5	0.25	4.80	-	0.94	0.29
5. Sand bed, cohesive bank and heavy sediment load*	3.08	0.37	9.70	-	-	0.29

* Sediment (mainly wash load) 2,000 to 3,000 ppm

It is possible to find local regime theories developed in every area where irrigation is practiced. The fact that these methods are not being transformed to other places is an indication that not all the physical parameters defining the problems are correlated by the regime methods (Raudkivi, 1990).

3.2.2   Tractive force theory

The tractive force methods are in use for boundary shear stress and sediment transport relationships. They use the concept of static stability of a canal in which there is no movement of material (both in the bed and side slopes). For a given design discharge, the canal dimensions and bed slope are determined considering the flow velocity not exceeding a permissible velocity or boundary shear stress not exceeding its critical value of bed and bank material. Under the tractive force theory two methods are in use:
−   maximum permissible velocity;
−   critical shear stress.

Method of maximum permissible velocity

If the adopted mean velocity of the canal is lower than the maximum permissible velocity, the canal is assumed to be stable. The maximum permissible velocity depends upon the bed material size and composition. Fortier and Scobey (1926) have proposed a set of such velocity limits. However, the problem with this approach is that if the water carries sediment then the canal may be stable at velocities higher than the limiting velocity. The problem of determining the permissible velocity still exists and hence requires a good judgement for each case on the part of the designer.

Many subsequent researchers have given methods to determine the maximum permissible velocities. Mavis *et al.* (1937) derived a formula to calculate the maximum permissible velocity at the bed (U_b) based on the size (d_{50} in m) and specific density (s) of the bed material. Where, U_b is in m/s measured at a distance *ks/30* from the bed (level of zero velocity) and *ks* is the Nikuradse's equivalent roughness height.

$$U_b = 3.283 d_{50}^{\frac{4}{9}} \sqrt{(s-1)} \qquad (3.13)$$

Carsten (1966), included also the bed slope in the determination of the permissible velocity. Neil (1967) has given design curves for uniform coarse material in terms of critical mean velocity. Yang (1973) used the conventional drag and lift concepts combined with the logarithmic velocity distribution and derived a formula to calculate the critical mean velocity.

$$V_c = \frac{2.5 w_s}{\log \dfrac{u_* d_{50}}{v} - 0.06} + 0.66 \qquad 1.15 < \frac{u_* d_{50}}{v} < 70 \qquad (3.14)$$

$$V_c = 2.05 w_s \qquad u_* d_{50}/v > 70 \qquad (3.15)$$

where

u_*	= shear velocity (m/s)
V_c	= critical mean velocity (m/s)
w_s	= particle fall velocity (m/s)
v	= kinematic viscosity of water (m²/s)

A summary of available equations for the prediction of velocity is given by Raudkivi (1990) and Simons and Senturk (1992). As indicated above the equations only provide guidance and judgement should be made based on the local conditions.

Method of critical shear stress

The critical shear stress is the limit after which the initiation of motion of the particles begins and the canal becomes unstable. Shear stress or drag force can be defined as the force that is resisted by friction force and, while in equilibrium, is equal and opposite in magnitude and direction. Shear stress (force per unit wetted area) is given by:

$$\tau_0 = \rho g R S \qquad (3.16)$$

where

τ_0	= tractive force per unit wetted area (N/m²)
ρ	= density of water (kg/m³)
g	= acceleration due to gravity (m/s²)

R = hydraulic radius (m)
S = bed slope

The distribution of shear stress in narrow and trapezoidal irrigation canals is non-uniform. In narrow irrigation canals the maximum shear stress is less than that predicted by equation (3.16) by some reduction factor. Lane (1953) determined experimentally that the adjustment factor for both the bed and side slopes largely depends on the width to depth ratio and side slope (Figure 3.1). Moreover, based on extensive work and field data of the U.S. Bureau of Reclamation prior to 1953, Lane (1955) established a critical tractive force diagram that relates the value with the mean diameter of the bed material for canals carrying water with different amounts of sediment.

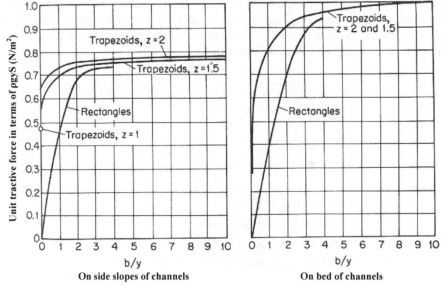

Figure 3.1   Maximum shear stress in a canal (Lane, 1953) (adapted from Chow (1983)).

These diagrams together with the uniform flow equations (Manning's or Chézy's) can be used to design a stable canal. Normally the grain size (d_{50}) of the bed material is known, so the calculation process involves assuming any three among side slope, bed slope, bed width, water depth and B-h ratio and finding the remaining two. The merits and limitations of this method have been discussed by Simons and Albertson (1963).

Straub's (1950) relation for critical bed slope can be used as a guide for the selection of the slope.

$$S_c = \frac{0.00025(d_{50} + 0.8)}{h} \qquad (3.17)$$

where

d_{50} = mean sediment diameter (mm)
h = water depth (m)

S_c = critical bed slope (m/m)

Similarly selection of the B-h ratio is another difficulty in the design of a canal using this method. Dahmen (1994) has proposed:

$$\frac{b}{h} = 1.76Q^{0.35} \qquad \text{for } Q > 0.2 \text{ m}^3/\text{s} \tag{3.18}$$

$$\frac{b}{h} = 1 \qquad \text{for } Q < 0.2 \text{ m}^3/\text{s} \tag{3.19}$$

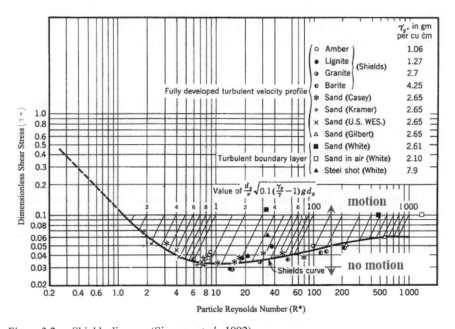

Figure 3.2 Shields diagram (Simons, *et al.*, 1992).

Apart from Lane's approach other theories that provide the condition for initiation of motion can be used for the design. Shield's relationship (1936) for the particle Reynolds number, Re$_*$ and the critical mobility parameter, θ_{cr} can also be used (Figure 3.2).

$$\text{Re}_* = \frac{u_* d_{50}}{v} \tag{3.20}$$

$$\theta_{cr} = \frac{\tau_{cr}}{(s-1)\rho g d_{50}} \tag{3.21}$$

where

d_{50} = representative bed material size (m)
Re.   = particle Reynolds number
s =specific density of bed material
θ_{cr} = critical mobility parameter
u_* = shear velocity (m/s)
τ_{cr} = critical shear stress (N/m^2)
v = kinematic viscosity (m^2/s)

Dahmen (1994) adopted a tractive force approach in combination with the bed width predictor (Etcheverry, 1915) and the related suspended sediment transport capacity parameter of De Vos to propose a canal design approach. This approach has been used in the modernization of one of the secondary canal systems of Sunsari Morang Irrigation Scheme (Department of Irrigation, 2003).

3.2.3   Rational theory

Three equations are required to determine the slope, depth, and width of a straight alluvial canal when the water and sediment discharges and bed material size are specified. The first two are provided by an alluvial resistance relation and a sediment transport equation. The third, a width relationship, can be obtained by optimization that is based on different hypotheses.

Chang (1980) used the concept of minimum stream power which states that for an alluvial channel, the necessary and sufficient condition of equilibrium occurs when the stream power per unit length is minimum subject to given constraints. The optimization is based on minimizing the bed slope according to given water discharge. The stream power (P_s) per unit channel length is given by:

$$P_s = \rho g Q S_f \tag{3.22}$$

where

g = acceleration due to gravity (m/s^2)
Q = flow rate (m^3/s)
S_f = slope of energy gradient
ρ = density of water (kg/m^3)

Equation (3.22) together with the suitable sediment transport and friction factor predictors make three equations that make it possible to solve for width, depth and slope of a stable regime canal.

Yang and Song (1979) used the concept of minimum unit stream power (Yang, 1972), which is defined as the time rate of potential energy expenditure per unit weight of water to explain the dynamic adjustments by natural rivers. An alluvial channel will adjust its velocity, slope, roughness and geometry and pattern in such a manner that a minimum amount of unit stream power is used to transport a given sediment and water discharge. The minimum unit stream power can be expressed as:

$$\frac{dy}{dt} = \frac{dx}{dt}\frac{dy}{dx} = VS_0 = a \quad \text{minimum} \tag{3.23}$$

where

S_0	= bed slope (m/m)
t	= time (s)
V	= average velocity (m/s)
x	= distance in the longitudinal direction (m)
y	= water surface elevation (m)
VS_0	= unit stream power (m/s)

Brebner and Wilson (1967) and Yang *et al.* (1981) used the concept of minimum energy dissipation in which they assumed that a system is in an equilibrium condition when its rate of energy dissipation is at a minimum value. The energy dissipation in a reach of stream length L is:

$$\Delta E = \gamma_w \left(Q + Q_s s \right) LS \tag{3.24}$$

where

ΔE	= energy loss per unit time (Nm/s)
Q	= water flow rate (m³/s)
Q_s	= sediment flow rate (m³/s)
s	= specific density of sediment
S	= bed slope of canal (m/m)
γ_w	= specific wieght of water (N/m³)

White *et al.*, (1982) proposed a design method based on the assumption that an alluvial canal adjusts its geometric characteristics and gradient in such a way that the sediment transporting capacity is maximized. Assuming the flow to be uniform and steady and the bed and bank material non-cohesive, they solved for bed width, water depth, flow velocity for maximum sediment transport capacity using the sediment predictor of Ackers and White (1973) and friction factor predictor of White *et al.*, (1980) (equation (3.25) to (3.29)). They have shown that maximization of the sediment transport capacity for fixed discharge and slope is equivalent to minimization of the friction slope for fixed discharge and sediment concentration, hence equivalent to the minimum stream power concept of Yang, and Song (1979) and Chang (1980).

$$V = \sqrt{32} \, \log(\frac{h}{d_{35}}) \left[\frac{F_{gr} \sqrt{g \, d_{35} \, (s-1)}}{u_*^{\,n}} \right]^{\frac{1}{1-n}} \tag{3.25}$$

For a given slope the shear velocity u_* is given by:

$$u_* = \sqrt{ghS_0} \tag{3.26}$$

the dimensionless mobility parameter F_{gr} is given by:

$$F_{gr} = (F_{fg} - A)\left[1.0 - 0.76\left(1 - \frac{1}{\exp{(\log D_*)}^{1.7}}\right)\right] + A \qquad (3.27)$$

the particle mobility parameter F_{fg} is given by:

$$F_{fg} = \frac{u_*}{\sqrt{g\,d_{35}\,(s-1)}} \qquad (3.28)$$

and the dimensionless grain size D_* is given by:

$$D_* = \left[\frac{(s-1)\,g}{v^2}\right]^{1/3} d_{35} \qquad (3.29)$$

where

A	= value of F_{gr} at the nominal, initial movement
d_{35}	= representative particle diameter (m)
D_*	= dimensionless grain parameter
F_{gr}	= dimensionless mobility parameter
g	= acceleration due to gravity (m/s^2)
h	= water depth (m)
n	= exponent
s	= specific density
u_*	= shear velocity (m/s)
V	= mean velocity (m/s)
v	= kinematic viscosity (m^2/s)

Wang *et al.* (1986) used 203 data sets from rivers and canals to test different external hypotheses and concluded that the assumption that the channels adjust to minimize their slope (equivalent maximizing sediment transport capacity) provided the best agreement with the data.

Sediment deposition can reduce the canal width while erosion can widen it. The extent that canals can widen depends on the strength of the bank material, a parameter which is difficult to specify. Because of these difficulties a regime relationship is sometimes used to provide the width equation. The sediment transport equations of Engelund and Hansen (1967), Ackers and White (1973), etc. could be combined with the friction equations (3.30 to 3.32) of Van Rijn (1984b, 1984a) or with the White, Paris and Bettess (1981b) friction equations ((3.25) to (3.29). Van Rijn's relations for the friction factor depending upon the flow regime can be written as :

$$C = 18 \log\left(\frac{12\,h}{3.3\,\dfrac{v}{u_*}}\right) \qquad \text{Smooth flow regime} \qquad (3.30)$$

$$C = 18 \log \left(\cfrac{12\,h}{k_s + 3.3\,\cfrac{v}{u*}} \right) \qquad \text{Transition flow regime} \qquad (3.31)$$

$$C = 18 \log \left(\frac{12\,h}{k_s} \right) \qquad \text{Rough flow regime} \qquad (3.32)$$

The canal width is determined from the Lacey regime equations, in which the wetted perimeter is a function of the square root of discharge. The second combination of methods was selected for the design of canals in Pakistan on the basis of comparison with a large set of Pakistan canal data (Bakker, *et al.*, 1986).

3.3   Water flow hydraulics

3.3.1   Governing equations

One dimensional unsteady flow in a river or a canal section (Figure 3.3) can be described by two dependent variables the water stage h and the discharge Q. Two equations are needed to find the state of the fluid flow that can be formulated from three physical laws: conservation of mass, energy and momentum (Cunge, *et al.*, 1980).

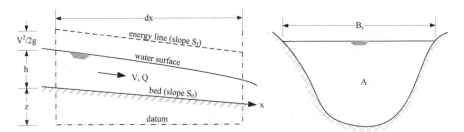

Figure 3.3   Definition sketch for water flow.

Conservation of mass and momentum, also known as continuity and dynamic equations for unsteady flows, given by de Saint-Venant are (Chow, 1983):

$$\frac{\partial h}{\partial t} + V\frac{\partial h}{\partial x} + h\frac{\partial V}{\partial x} = 0 \qquad (3.33)$$

$$\frac{\partial V}{\partial t} + V\frac{\partial V}{\partial x} + g\frac{\partial h}{\partial x} + g\frac{\partial z}{\partial x} = -gS_f \qquad (3.34)$$

For steady and gradually varied flow these equations reduce to:

$$\frac{\partial (Vh)}{\partial x} = 0 \qquad \therefore q = \text{constant} \tag{3.35}$$

$$\frac{dh}{dx} = \frac{\left(S_0 - S_f\right)}{\left(1 - Fr^2\right)} \tag{3.36}$$

with Froude number: $$Fr = \sqrt{\frac{Q^2 B_s}{gA^3}} \tag{3.37}$$

where

A	= area of water flow (m^2)
B_s	= water surface width (m)
g	= acceleration due to gravity (m/s^2)
h	= water depth (m)
Q	= flow rate (m^3/s)
S_0	= bed slope (m/m)
S_f	= friction slope (m/m)
V	= flow velocity (m/s)
z	= bed level above datum (m)

The assumptions made in deriving the above equations are (Cunge, *et al.*, 1980):
- the flow is one dimensional, the flow velocity is uniform over the cross section and the water level across the section is horizontal;
- the streamline curvature is small and vertical accelerations are negligible, hence the pressure is hydrostatic;
- the effect of boundary friction and turbulence can be accounted for through resistance laws analogous to those used for steady state flow;
- the average channel bed slope is small, so that the cosine of the angle it makes with the horizontal may be replaced by unity.

In case of uniform flow, the water depth along the flow remains constant. So equation (3.36) reduces to:

$$\frac{dh}{dx} = 0 \qquad \Rightarrow S_0 = S_f \tag{3.38}$$

Most common uniform flow formulae are those of Chézy and Manning:

$$V = C\sqrt{RS_f} \tag{3.39}$$

$$V = \frac{1}{n} R^{\frac{2}{3}} S_f^{\frac{1}{2}} \tag{3.40}$$

where

C = Chézy's roughness coefficient ($m^{1/2}/s$)
n = Manning's roughness coefficient ($s/m^{1/3}$)
R = hydraulic radius (m)
S_f = friction slope and is equal to bed slope in uniform flow

These are the basic hydraulic equations used in the canal design. After the evaluation of the design discharge, equations (3.39) or (3.40) are used to derive the geometry of the canal, which is based on the assumption that the flow is steady and uniform. The roughness factor C is selected on the basis of the bed material, proposed maintenance level of the canal bed width to water depth ratio, etc. Selection of the accurate roughness coefficient is crucial in the success of the scheme as it has direct effect on the conveyance and water level of the canal. Besides, most of the time the flow in the canal is less than the design value. To maintain the desired water level and to divert the targeted amount to the laterals flow control structures are provided that make the flow non-uniform. Again if the proposed maintenance level is not met the roughness of the section will be more than the design value (smoothness less) and less conveyance for the same depth results.

3.3.2   Roughness prediction

The total flow rate in an open canal with single roughness along its perimeter can be computed from:

$$Q = CA\sqrt{RS_f} \qquad\qquad (3.41)$$

where Chézy's roughness coefficient is given by:

$$C = 18\log\left(\frac{12R}{k_s}\right) \qquad\qquad (3.42)$$

If the roughness height (k_s) is not constant throughout the perimeter then an equivalent roughness height (k_{se}) has to be found to determine the flow parameter. Different methods are available to determine equivalent roughness of a section. Yen (2002) discussed in detail the different aspects of computing composite roughness of an open canal.

3.4   Sediment transport aspects

The sediment transport process, available predictors for estimating sediment transport rates under equilibrium and non-equilibrium conditions, their strengths and limitations will be discussed in this section. The understanding of this aspect will help to formulate the canal design procedure for sediment transport.

3.4.1   Initiation of motion

Under the action of flow the sediment in the bed starts to move after a certain critical condition. Three hydraulic parameters that can be used to define the critical condition for the incipient motion are: shear stress (drag force), average velocity and flow power. Among the theories based on critical shear stress, the Shields' diagram is the most widely used criteria to describe the condition for initiation of motion for non-cohesive and uniform sediment on a horizontal bed. Shields' formula for drag force is given by:

$$\theta_{cr} = \frac{u_*^2}{(s-1)gd_{50}} = \frac{\tau_{cr}}{(s-1)\rho gd_{50}} = f\left(\frac{u_{*,cr}d_{50}}{\nu}\right) \tag{3.43}$$

where

d_{50} = representative bed material size (m)

s = specific density of bed material

θ_{cr} = critical mobility parameter

u_* = shear velocity (m/s)

τ_{cr} = critical shear stress (N/m²)

ν = kinematic viscosity (m²/s)

Equation (3.43) shows that when a grain starts to move, the ratio of the drag force acting on a grain to its weight is a function of the grain Reynolds number (Re_*) (equation 3.20). With the help of experiments using sediments of different specific gravities, Shield obtained the value of the function f in equation (3.43) and plotted the results in graphical form what is popularly known as the Shield's diagram (Figure 3.2).

The solution using Shields' diagram requires a trial and error procedure, since the parameter u_* occurs in both the axes of the diagram. Yalin's (1977) equation can be used to eliminate the trial and error procedure.

$$D_* = \frac{R_{e*}^2}{\theta} = \left[\frac{(s-1)g}{\nu^2}\right]^{\frac{1}{3}} d_{50} \tag{3.44}$$

Once D* is known Van Rijn (1993) relationships can be used to determine the value of θ_{cr} and solve equation (3.43) for τ_{cr} .

$$\theta_{cr} = 0.24D_*^{-1} \qquad \text{for } 1 < D_* \leq 4 \tag{3.45}$$

$$\theta_{cr} = 0.14D_*^{-0.64} \qquad \text{for } 4 < D_* \leq 10 \tag{3.46}$$

$$\theta_{cr} = 0.04D_*^{-0.1} \qquad \text{for } 10 < D_* \leq 20 \tag{3.47}$$

where

D_* = dimensionless grain parameter

s =specific density of sediment

τ_{cr} = shear stress (N/m^2)

$u_{*,cr}$ = critical shear velocity (m/s)

v = kinematic viscosity of water (m^2/s)

In the second method, the critical velocity for the initiation of motion (ref. section 3.2.2) can be derived when a logarithmic or an exponential velocity formula is substituted in equation (3.43).

Similarly the third method relates the critical value of flow power to put the sediment particles in motion. Although the drag force, velocity and flow power for incipient motion provide three different expressions for the same phenomenon and can be mutually transformed from one to another, they still represent three different approaches based upon three different concepts. Considering the effect of turbulence on the incipient motion of sediment, the drag force is a better index than the bed velocity (Chien and Wan, 1999). American Society of Civil Engineers recommends the use of critical shear stress wherever possible for the calculation of initiation of motion (Task Committee on Preparation of Sedimentation Manual, 1966a).

3.4.2 Sediment transport under equilibrium

Sediment transport is assumed to be in equilibrium when there is no net change of total sediment load (bed and suspended) with time and space. By equilibrium condition it is assumed that the amount of sediment transported by the flowing water is equal to its transporting capacity without any restriction of sediment exchange in the bed. Even during equilibrium condition, there is a constant exchange of sediment between bed and flowing water but the net exchange is zero.

Bed-load transport

The transport of bed material particles by a flow of water can be in the form of bed-load and suspended load, depending on the size of the bed material particles and the flow conditions. Although in natural conditions there is no sharp division between the bed-load transport and the suspended load transport, it is necessary to define a layer with bed-load transport for mathematical representation. Bagnold (1966) defines bed-load as the particles which are in successive contacts with the bed and the processes are governed by gravity. The particles in bed-load move by saltation, rolling or sliding in the bed layer. Suspended load is defined as the sediment that is lifted by the upward components of turbulent currents and that stays in suspension for an appreciable length of time (Simons, *et al.*, 1992).

Bed-load transport process and the formula to compute the yield has been given by many scientists. The simplest yet fairly reliable empirical equation based on experimental data is by Meyer-Peter (Raudkivi, 1990). Einstein (1950) introduced statistical methods to represent the turbulent behaviour of the flow and gave a bed-load function. Bagnold (1966) equation is based on a physical concept and mechanical analysis. He introduced an energy concept and related the sediment transport rate to the work done by the fluid. Engelund (1966), Ackers and White (1973) and Yalin (1977) equations are mainly based on Einstein or Bagnold concepts but deduced using dimensional analysis. Van Rijn (1984a) solved the equations of motion of an individual bed-load particle and computed the saltation characteristics and the particle velocity as a function of the flow conditions and the particle diameter for plane bed conditions.

Suspended load transport

Transport of sediment takes place in suspended mode, when the bed-shear velocity (u_*) exceeds the particle fall velocity (w_s). The particles can be lifted to a level at which the upward turbulent forces will be comparable to or higher than the submerged particle weight and as a result the contact of the particle with the bed in the suspension mode is occasional and random. The particle velocity in longitudinal direction is almost equal to the fluid velocity. Usually, the behaviour of the suspended sediment particles is described in terms of sediment concentration, which is the solid volume (m^3) per unit fluid volume (m^3) or the solid mass (kg) per unit fluid volume (m^3). The principle feature that distinguishes suspended sediment transport from bed-load transport is the time taken for the suspension to adapt to changes in flow conditions (Galappatti, 1983).

According to Bagnold (1966) suspension will occur for the bed shear velocity (u_*) equal or larger than the particle fall velocity (w_s), while Van Rijn (1984b) argues that suspension will start at considerably smaller bed-shear velocities. From his experimental data Van Rijn (1984b) proposed the following conditions for the initiation of suspension:

$$\frac{u_{*,cr}}{w_s} = 4D_*^{-1} \quad \text{for} \quad 1 < D_* \leq 10 \tag{3.48}$$

$$\frac{u_{*,cr}}{w_s} = 0.4 \quad \text{for} \quad D_* > 10 \tag{3.49}$$

Discharge of the sediment per unit width can be written as:

$$q_{s,c} = \int_a^h ucdz \tag{3.50}$$

$$q_{s,c} = c_a \bar{u} \int_a^h \frac{u}{\bar{u}} \frac{c}{c_a} dz \tag{3.51}$$

where
c   = sediment concentration at height z from the bed
c_a = reference concentration at reference height a (z = a) above bed level
h   = water depth (m)
$q_{s,c}$ = volumetric suspended transport (m^2/s)
u   = velocity at height z above the bed (m/s)
$\bar{u}$ = depth averaged flow velocity (m/s)

For the solution of equation (3.51), the velocity profile, concentration profile and concentration at reference level should be known. For steady and uniform conditions of water and sediment, the concentration profile of sediment in the vertical is in equilibrium. A number of relations are available for the prediction of suspended sediment transport rates. Some of them are based on analytical approaches but still need experimental results to derive certain parameters. Mostly two methods are in use to determine the sediment concentration profile: the diffusion methods and the energy or gravitational methods.

Diffusion method

The diffusion models are based on the mass balance equation. The partial differential equation that governs the transport of suspended sediment by convection and turbulent diffusion under gravity is (Galappatti, 1983):

$$\frac{\partial c}{\partial t}+v\frac{\partial c}{\partial y}+u\frac{\partial c}{\partial x}+w\frac{\partial c}{\partial z}=w_s\frac{\partial c}{\partial z}+\frac{\partial}{\partial x}\left(\epsilon_x\frac{\partial c}{\partial x}\right)+\frac{\partial}{\partial y}\left(\epsilon_y\frac{\partial c}{\partial y}\right)+\frac{\partial}{\partial z}\left(\epsilon_z\frac{\partial c}{\partial z}\right) \quad (3.52)$$

where

c	= suspended sediment concentration (m^3 sediment/m^3 water)
t	= time coordinate (s)
w_s	= fall velocity (m/s)
x, y, z	= length coordinates (m)
$\varepsilon_x, \varepsilon_y, \varepsilon_z$	= sediment mixing coefficients in x, y and z direction (m^2/s)
u, v, w	= velocity components in x, y and z directions (m/s)

For a two-dimensional flow in the vertical plane, equation (3.52) reduces to:

$$\underbrace{\frac{\partial c}{\partial t}+u\frac{\partial c}{\partial x}+w\frac{\partial c}{\partial z}}_{\text{Convection terms}}=\underbrace{w_s\frac{\partial c}{\partial z}+\frac{\partial}{\partial x}(\varepsilon_x\frac{\partial c}{\partial x})+\frac{\partial}{\partial z}(\varepsilon_z\frac{\partial c}{\partial z})}_{\text{Diffusion terms}} \quad (3.53)$$

Convection terms   Diffusion terms

A numerical solution of the equation is possible when the velocity components, the fall velocity and the mixing coefficients are known. If the sediment mixing coefficient in the direction of flow (ε_x) is negligible as compared to the vertical one (ε_z), i.e.:

$$\frac{\partial}{\partial x}(\varepsilon_x\frac{\partial c}{\partial x})=0 \quad (3.54)$$

the equation (3.53) for a one-dimensional problem reduces to:

$$\frac{\partial c}{\partial t}+u\frac{\partial c}{\partial x}=w_s\frac{\partial c}{\partial z}+\frac{\partial}{\partial z}(\varepsilon_z\frac{\partial c}{\partial z}) \quad (3.55)$$

If the vertical concentration profile of suspended sediment is in equilibrium, then the sediment diffusion process is steady and uniform; hence all the derivatives with respect to distance (x) and time (t) in equation (3.55) are equal to zero (Chien and Wan, 1999). Then the equation can be written as:

$$w_s \frac{\partial c}{\partial z} + \frac{\partial}{\partial z}(\varepsilon_z \frac{\partial c}{\partial z}) = 0 \qquad (3.56)$$

Integration of equation (3.56), assuming the diffusion coefficient for sediment ($\varepsilon_z = \varepsilon_s$) to be constant over the depth gives:

$$\frac{c}{c_a} = \exp\left[-\frac{w_s(h-a)}{\varepsilon_s}\right] \qquad (3.57)$$

where

a	= reference level (m)
c	= concentration at depth z (ppm)
c_a	= reference concentration (ppm)
h	= flow depth (m)

If $\varepsilon_s \cong \varepsilon$ (momentum diffusion of water) then:

$$\varepsilon_s \cong \kappa u_* z(1 - z/h) \qquad (3.58)$$

Substituting the value of ε_s in equation (3.58), for $w/u_* \ll 1$, the concentration will become nearly uniform. Moreover if the diffusion coefficient for sediment is assumed to be equal to momentum diffusion coefficient of the flow, integration of equation (3.57) reduces to the Rouse equation (Raudkivi, 1990), which is given by:

$$\frac{c}{c_a} = \left[\left[\frac{h-z}{h-a}\right]\frac{a}{z}\right]^{\frac{w_s}{\kappa u_*}} \qquad (3.59)$$

In deriving (3.59) the velocity distribution was represented by a logarithmic equation, which is invalid near the bed and also, inaccurate near the water surface. As the sediment concentration is maximum near the bed and has a great effect on the mean sediment concentration the use of equation (3.59) is inaccurate (Chiu, et al., 2000).

Energy concept

An energy approach to the suspension problem, known as the gravitational theory, was proposed by Velikanov (Raudkivi, 1990). It has been demonstrated that the actual differences between the gravitational theory and the diffusion method are small and rest on the differences in assumptions made. The equation from the energy theory is as given:

$$w_s \overline{c} + \frac{\overline{u\tau}}{\rho g(s-1)}\frac{\partial \overline{c}}{\partial z} = 0 \qquad (3.60)$$

where

$\overline{c}$ = time averaged concentration (m^3/m^3)

s = specific density of sediment particle

$\overline{u}$ = time averaged velocity (m/s)

w_s = particle fall velocity (m/s)

z = vertical depth (m)

$\overline{\tau}$ = time averaged shear stress (N/m^2)

ρ = density of water (kg/m^3)

This equation is of the same form as obtained from the diffusion theory, except that the exchange coefficient has a different form. Using constant shear stress and logarithmic velocity distribution, the concentration profile is given as (Chien and Wan, 1999):

$$\frac{c}{c_a} = e^{-\beta\zeta} \tag{3.61}$$

$$\zeta(\eta,\alpha) = \int_{\eta_a}^{\eta} \frac{d\eta}{(1-\eta)\ln[1+\eta/\alpha]} \tag{3.62}$$

$$\beta = (s-1)\frac{\kappa w_s}{S_0\sqrt{ghS_0}} \tag{3.63}$$

$$\eta = \frac{z}{h} \quad \text{and} \quad \alpha = \frac{a}{h} \tag{3.64}$$

where

c = concentration at depth z (ppm)

c_a = reference concentration (ppm) at reference height 'a' from bed

h = water depth (m)

S_0 = bed slope

s = specific density of sediment particle

w_s = particle fall velocity (m/s)

κ = von Karman constant

In both diffusion and energy approaches, the major challenges are to determine the reference height and reference concentration. The concentration at any point above the bed can be determined only with the known concentration at reference level, which cannot be determined theoretically.

Based on the above mentioned two concepts, different sediment transport predictors have been derived using theoretical and experimental results. Some of them have a theoretical basis, but the derivations are oversimplified. Hence the relationships are more empirical in nature. There are various methods available to calculate the sediment transport rate for a large range of flow conditions and sediment characteristics. Some of them calculate the bed-load and suspended load separately and add them to get the total load, while others directly give the total load. In the first category theories forwarded by Einstein (1950), Bagnold (1966), Toffaletti (1969), Van Rijn (1984a, 1984b) can be mentioned. In the second category Colby (1964), Bishop *et al.* (1965), Engelund and Hansen (1967), Ackers and White (1973), Yang (1973) and Brownlie (1983) are some of the methods that give the total sediment transport. A brief description of the methods that compute total load is given in the following paragraphs.

Engelund and Hansen method. The Engelund and Hansen method (1967) is based on the energy balance concept. The total sediment transport function is calculated by:

$$\phi = \frac{q_s}{\sqrt{(s-1)\, g\, d_{50}^3}} \tag{3.65}$$

$$\theta = \frac{u_*^2}{(s-1)\, g\, d_{50}} \tag{3.66}$$

$$\phi = \frac{0.1\,\theta^{2.5}\, C^2}{2\, g} \tag{3.67}$$

$$q_s = \frac{0.05\, V^5}{(s-1)^2\, g^{0.5}\, d_{50}\, C^3} \tag{3.68}$$

where

C	= Chézy's coefficient ($m^{1/2}/s$)
d_{50}	= representative particle size (m)
q_s	= volumetric total sediment transport (m^2/s)
s	= specific density (density of particle/density of water)
V	= depth averaged velocity (m/s)
θ	= dimensionless mobility parameter
ϕ	= dimensionless transport parameter

Ackers and White. Ackers and White (1973) derived an empirical formula based on 925 sets of flume and field data. In establishing the formula for sediment transport capacity, Ackers and White adopted the energy concept that the intensity of sediment transport is related to the power provided by the flow and they assumed the efficiency of sediment transport to be proportional to a dimensionless mobility parameter. This method describes the sediment transport in terms of dimensionless parameters: D* (grain size sediment parameter), F_{gr} (mobility parameter) and G_{gr} (transport parameter). These parameters for D_* in the range of 1 to 60 are given as:

$$D_* = [\frac{(s-1)\,g}{v^2}]^{1/3}\,d_{35} \tag{3.69}$$

$$F_{gr} = \frac{u_*^{\,n}}{\sqrt{g\,d_{35}\,(s-1)}}[\frac{V}{\sqrt{32}\,\log(\frac{10\,h}{d_{35}})}]^{1-n} \tag{3.70}$$

$$n = 1.00 - 0.56\,\log D_* \tag{3.71}$$

$$G_{gr} = c\,(\frac{F_{gr}}{A} - 1)^m \tag{3.72}$$

$$A = \frac{0.23}{\sqrt{D_*}} + 0.14 \tag{3.73}$$

$$m = \frac{9.66}{D_*} + 1.334 \tag{3.74}$$

$$\log c = 2.86\,\log D_* - (\log D_*)^2 - 3.53 \tag{3.75}$$

The Ackers and White function to determine the total sediment transport reads as:

$$q_s = G_{gr}\,V\,d_{35}\,(\frac{V}{u_*})^n \tag{3.76}$$

where
A	= value of F_{gr} at the nominal, initial movement
c	= coefficient in the transport parameter G_{gr}
D_*	= dimensionless grain parameter
d_{35}	= representative particle diameter (m)
F_{gr}	= dimensionless mobility parameter
g	= acceleration due to gravity (m/s^2)
G_{gr}	= dimensionless transport parameter
h	= water depth (m)
m	= exponent in the transport parameter G_{gr}
n	= exponent in the dimensionless mobility parameter F_{gr}
q_s	= total sediment transport per unit width (m^2/s)
s	=specific density
u_*	= shear velocity (m/s)
v	= kinematic viscosity (m^2/s)
V	= mean velocity (m/s)

The c and m coefficients were later modified (HR Wallingford, 1990, Ackers, 1993). These modifications were necessary as the original relations predicted transport rates which were considered too large for relatively fine sediment (d_{50} <

0.2 mm) and for relatively coarse sediment. The modified coefficients for a dimensionless grain parameter (D_*) in the range of 1 to 60 are:

$$\log c = 2.79 \log(D_*) - 0.98(\log D_*)^2 - 3.46 \tag{3.77}$$

$$m = \frac{6.83}{D_*} + 1.67 \tag{3.78}$$

Brownlie method. Brownlie's method (1981) to compute the sediment transport is based on a dimensional analysis and calibration of a wide range of field and laboratory data, where uniform conditions prevailed. The transport (in ppm by weight) is calculated by:

$$q_s = 727.6 \, c_f \, (F_g - F_{gcr})^{1.978} \, S^{0.6601} \left(\frac{R}{d_{50}}\right)^{-0.3301} \tag{3.79}$$

where, F_g is grain Froude number and is given by:

$$F_g = \frac{V}{[(s-1) \, g \, d_{50}]^{0.5}} \tag{3.80}$$

F_{gcr} is critical grain Froude number and is given by:

$$F_{gcr} = 4.596 \, \tau_{*_0}^{0.5293} \, S^{-0.1405} \, \sigma_s^{-0.1696} \tag{3.81}$$

critical dimensionless shear stress (τ_{*_0}) is given by:

$$\tau_{*_o} = 0.22 \, Y + 0.06 \, (10)^{-7.7 \, Y} \tag{3.82}$$

the coefficient Y is given by:

$$Y = (\sqrt{s-1} \, R_g)^{-0.6} \tag{3.83}$$

R_g is the grain Reynold number and is calculated from:

$$R_g = \frac{(g \, d_{50}^3)^{0.5}}{31620 \, v} \tag{3.84}$$

where

    c_f = coefficient for the transport rate (c_f is 1 for laboratory conditions and 1.268 for field conditions)

    d_{50} = median diameter (mm)

    F_g = grain Froude number

    F_{gcr} = critical grain Froude number

g = gravity acceleration (m/s^2)

R_g = grain Reynolds number

R = hydraulic radius (m)

S = bed slope

s = specific density

v = kinematic viscosity (m^2/s)

τ_{*0} = critical dimensionless shear stress

σ_s = geometric standard deviation

3.4.3   Non-equilibrium sediment transport

Non-equilibrium sediment transport is a condition when the concentration varies along the flow direction. Non-equilibrium conditions may be found both in uniform and non-uniform flow conditions. Continuous entrainment of sediment downstream of a reservoir, settling basin or flow control structure and deposition in a settling basin are examples of non-equilibrium sediment transport under uniform flow conditions. In non-uniform flow, the flow parameters vary along the canal length and so the sediment transport capacity will also change. This will result in a continuous readjustment of the sediment concentration in flowing water. Readjustment takes place by continuous deposition or entrainment of sediment to or from the canal bed. The adjustment process is almost instantaneous in case of bed-load, since this mode of transport takes place close to the bed, while the adjustment is relatively slow in case of suspended load transport because it takes time and hence distance for the particles to settle down from suspension or to be mixed into the flow from the bed, depending on the ratio of the bed shear velocity and fall velocity. This lag in the adjustment with changing flow creates a non-equilibrium pattern of sediment transport. For solving the non-equilibrium concentration profile, usually the convection diffusion equation is used (ref. subsection 3.4.2).

3.5   Modelling aspects

Sediment transport modelling includes two processes; the solution of the hydraulic parameters and the sediment mixing and transport parameters. Hence two models are required, namely the flow model and the sediment transport and morphological model. Depending upon whether the computation is for a single time step or multi time step, the combined model may be divided into two types (Van Rijn, 1993):

-   initial or sediment transport model that computes the sediment transport rate and bed level change for one time step;
-   dynamic morphological model, which computes the flow velocity, the sediment transport rates, the bed level change and again the new flow velocity and the process is continued for the next time step.

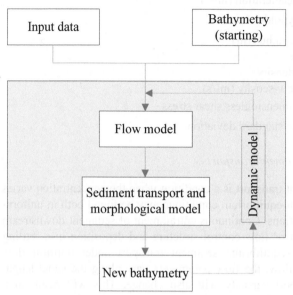

Figure 3.4   Structure of initial and dynamic models (after Van Rijn, (1993)).

Depending upon the number of independent space variables used to define the water flow and sediment movement processes, the models can be (Van Rijn, 1993):
– three-dimensional (3D);
– two-dimensional vertical (2DV)
– two-dimensional horizontal (2DH);
– one-dimensional (1D).

Three-dimensional models (3D)

The 3D models are used for complex flow patterns and usually when small scale and short-term processes are to be investigated. Basically, two types of models are in use:
– *the depth-integrated approach.* The vertical dimension is eliminated by means of an asymptotic solution as introduced by Galappatti and Vreugdenhil (1985). This type of models can be used in case of a canal geometry and where the difference between the local true suspended sediment transport and the local equilibrium transport is relatively small (Wang, 1989), otherwise the results are not accurate;
– *the three dimensional approach.* In this approach a full unsteady three-dimensional model for fluid velocities and sediment concentrations is used. Examples of this approach are the models of Sheng and Butler (1982), Miller (1983), Wang and Adeff (1986), Wang, *et al.* (1986), McAnally, *et al.* (1986), O'Connor and Nicholson (1988), Van Rijn and Meijer (1988), Lin and Falconer (1996) and others. To operate a 3D-model, the flow velocities and mixing coefficients must be known a priori. These models are useful for a complicated geometry with flow separation conditions and where the simulations are for short periods. The results of 3D models are used to evaluate or calibrate 1D or 2D models.

Two-dimensional vertical models (2DV)

The flow and sediment parameters are averaged over the width in 2DV models. These models are useful to predict transport rates, sedimentation and erosion in rivers, estuaries and coastal waters and in sediment traps in irrigation canals (Van Rijn, 1993).

Smith and O'Connor (1977) presented a two dimensional vertical model based on the laterally-integrated momentum and continuity equations for the fluid and sediment phases, and a "one-equation" turbulence closure model to represent the fluid shear stresses and diffusion coefficients. Celik and Rodi (1984, , 1988) presented a similar model as that of Smith and O'Connor. However, the former applied a "two-equation" turbulence closure model (K-Epsilon model). Bechteler and Schrimpf (1984) presented a relatively simple two-dimensional model neglecting vertical convection and horizontal diffusion. The flow field was represented by logarithmic velocity profiles. Kerssens and Van Rijn (1977), Kerssens, et al. (1979) presented a two-dimensional vertical model for gradually varying flows neglecting vertical convection and horizontal diffusion. Logarithmic velocity profiles were used to represent the fluid velocity field. The vertical sediment mixing coefficients were represented by a parabolic-constant distribution. A concentration-type boundary condition was applied at the bed, assuming instantaneous adjustment to equilibrium conditions close to the bed. Later, the flow width was introduced to extend the applicability of the model to gradually varying flows in transverse direction. A method based on the application of flexible profiles (shape functions) was introduced to obtain a better description of the velocity profiles and the sediment mixing coefficients (Van Rijn, 1986, , 1987).

Two-dimensional horizontal models (2DH)

Two dimensional horizontal models are based on depth integrated equations of motion in combinations with a wave model and a sediment transport model (Van Rijn, 1993). Assuming the velocity and sediment concentration profile similar to that at equilibrium conditions, Galappatti and Vreugdenhi (1985) proposed a depth integrated model using a convection-diffusion equation. Ziegler and Nisbet (1994) developed a model to simulate the suspended transport of fine-grained sediment, both cohesive and non-cohesive. They used the previously developed SEDZEL model (Ziegler and Lick, 1986, Gailani, et al., 1991) and modified the framework to accommodate the simulation of non-cohesive sediment part also.

Olsen (1999) developed a two-dimensional numerical model for the simulation of flushing of sediments from reservoirs. It solves the depth-averaged Navier-Stokes equations on a two-dimensional grid. A zero-equation turbulence model is used. The resulting flow field is extrapolated to three dimensions, and the convection-diffusion equation for the sediment concentration is solved. A formula for the bed concentration is used as boundary condition, resulting in a calculation of the bed material load.

García-Martínez, et al. (1999) proposed a relatively simple two-dimensional finite element mathematical model to simulate suspended sediment transport in coastal regions. Sediment transport is formulated in terms of a hydrodynamic model plus a convection-diffusion equation with source and sinks terms representing the erosion and deposition processes.

Guo and Jin (2002) proposed a two-dimensional mathematical model for non-uniform suspended sediment transport to simulate riverbed deformation. The problem of exchange of sediment between suspended and bed material is handled by dividing the mixture into several size groups in which it is assumed to be uniform.

One dimensional model (1D)

One-dimensional models are most frequently used to simulate the large-scale morphological changes in rivers and estuaries. The sediment transport and the bed roughness can be represented by simple formulas in terms of the mean flow variables and sediment properties. Analytical solutions can be obtained for simple schematized cases (De Vries, 1975). Numerical solutions are required for more realistic cases. If required, the non-uniformity of the bed material can be taken into account by dividing the bed material in a number of size classes (Ribberink, 1987, Armanini and Di Silvio, 1988). Adjustment effects related to the suspended sediment transport process can be taken into account by using the depth-integrated approach (Galappatti, *et al.*, 1985).

Detailed information of one-dimensional river models with respect to numerical solution methods and practical applications is given by Cunge, *et al.* (1980) and by Jansen, *et al.* (1979). A state of the art review is given by de Vries, *et al.* (1989).

Lin, *et al.* (1983) used an expression for the rate of bed changes along with the momentum and continuity equations of water and sediment to compute the sediment transport rate. Guo and Jin (1999) presented a one-dimensional mathematical model to calculate the bed variations in alluvial canals based on the depth-averaged and moment equations for unsteady flow and sediment transport in open canals. The momentum equation for suspended sediment transport is derived by assuming a simple vertical distribution for suspended sediment concentration. By introducing the sediment-carrying capacity, the suspended sediment concentration can be solved directly from the sediment transport and its moment equations. Differential equations are then solved by using the control-volume formulation.

3.6   Sediment transport models

As discussed above the water flow and sediment transport processes can be expressed in mathematical relationships and solved by using different techniques. It should ,however, be realized that a complete mathematical representation of the real process is not possible and they incorporate only the important parameters discarding the secondary ones that are thought to be less important in view of the purpose of investigation. Different models have been developed and are in use with varying assumptions and capability.

Models that can simulate the sediment transport process are mainly intended for the rivers. They were generally not developed with specific characteristics of irrigation systems in mind. Some of them offer possibilities of running user written algorithms (Clemmens, *et al.*, 2005), but still there are limitations due to the functional and computational environment of the model. There are few models like Design Of Regime Channels (DORC) (HR Wallingford, 1992), ISIS (Halcrow, 2003), Simulation of Irrigation Canal (SIC) (Malaterre, 2007), etc. that have been customized for irrigation canals, but they do not take into account the specific difference the irrigation canals possess, in terms of width to depth ratio and finite but different roughness of side slopes. Méndez (1998) has reviewed the different models and their capability to simulate the sediment transport process especially in irrigation canals. Some of the models that take care of the sediment transport process are briefly described below.

CanalCAD (Holly and Parrish, 1992) simulates both steady and unsteady flow in canal systems with manual or automatic gates. Hydrodynamic channel computations are based on the full, dynamic, one-dimensional de St. Venant equations and utilize the Preissman four-point implicit finite-difference scheme. It is designed for the design, analysis, and operation of irrigation canals comprising sub-critical flow in a single in-line system of pools and appurtenant structures including turnouts, in-line weirs, check structures, culverts, off-line storage reservoirs, etc.

SHARC (Sediment and Hydraulic Analysis for Rehabilitation of Canals) is developed by Hydraulic Research Wallingford (HR Wallingford, 2002). It integrates a sediment routing model with design packages for sediment control structures and other modules. For routing sediment transport in a canal, the model includes DORC (Design of Regime Canals) software. DORC includes regime methods, tractive force, rational methods, Manning, alluvial friction predictors, and sediment transport for sand and silt (HR Wallingford, 1992). The regime methods used in the model are the Lacey and Simons and Albertson methods. Among the rational methods the White, Paris and Bettess (1982) and the Chang (1985) method are included. The model is recommended for predicting the transport capacity of canals, but it is not possible to determine the sediment transport under non-equilibrium conditions (HR Wallingford, 1991).

DUFLOW is a micro-computer program for simulating one-dimensional unsteady flow in open canals. It was jointly developed by Rijkswaterstaat, Delft University of Technology (TUD) and the International Institute for Hydraulic Infrastructure and Environment (IHE) (Spaans, 2000). The program was initially developed for simple canal networks with simple structures. A module for calculating the suspended sediment transport in non-equilibrium conditions was incorporated. A depth integrated model based on the advection-diffusion equation is used for the transport model. The model is driven by the deposition and re-suspension fluxes at the bed so that the morphological changes over a small time interval can be computed (Clemmens, *et al.*, 1993).

MIKE 11 (Danish Hydraulic Institute, 1993) is developed by Danish Hydraulic Institute (DHI). It is a one-dimensional hydrodynamic software package including a full solution of the de St. Venant equations, plus many process modules for advection-dispersion, water quality and ecology, sediment transport, rainfall-runoff, flood forecasting, real-time operations, and dam break modelling (Danish Hydraulic Institute, 2002). The sediment transport modules of MIKE 11 are applied for:
- long-term morphological changes in the river;
- erosion, transport and deposition of contaminated and polluting sediments;
- optimization of capital and maintenance of dredging for e.g. for navigation;
- sediment management in rivers and reservoirs;
- river restoration projects.

MIKE 21C is an integrated river morphology modelling tool based on a curvilinear version of the water model MIKE 21 and adjusted to river applications

(Danish Hydraulic Institute, 2009). The model can be used to simulate changes in the river bed and planform, including bank erosion, scouring, shoaling associated with, for instance, construction works and changes in the hydraulic regime. The model system simulates transport of all sediment sizes from fine cohesive material to gravel using a multi-fraction approach and a number of transport formulae. Possibility of selecting fixed layer or variable layer. In case of fixed layer option the re-suspension is unlimited.

SOBEK is capable of handling one-dimensional problems in open canal networks. Apart from steady or unsteady flow, the model can touch various other physical processes like salt intrusion and morphology, sediment transport and water quality. The model was developed by Delft Hydraulics and the Institute for Inland Water Management and Waste Water Treatment of the Netherlands (Delft Hydraulics and Ministry of Transport Public Works and Water Management, 1994a). Applications of SOBEK are mentioned for river training, dredging optimization, river bed cut-offs, water quality, water flow for industry, drinking water, cooling water and irrigation, regime changes of rivers, flood risk, low water, etc. Sediment transport and morphology modules are mainly used for the indication of aggradation and degradation of river reaches due to river bend cut-offs, dredging, river training, water extraction, reservoir operation or flooding. The transport of sand in rivers and estuaries is estimated by using one of the following sediment transport formulae: Ackers and White, Engelund and Hansen, Meyer-Peter and Muller, Van Rijn, Parker and Klingeman and a general user-adjustable formula.

SIC (Simulation of Irrigation Canals) was developed by Cemagref, France (Malaterre, *et al.*, 1997). The SIC (Simulation of Irrigation Canals) software is a mathematical model which can simulate the hydraulic behaviour of most of the irrigation canals or rivers under steady and unsteady flow conditions. The main purposes of the model are (Malaterre, 2007):
- to provide a research tool to gain an in-depth knowledge of the hydraulic behaviour of the main and secondary canals;
- to identify, through the model, appropriate operational practices at regulating structures with a view to improve the present canal operations;
- to evaluate the influence of possible modifications to some design parameters with a view to improve and maintain the capacity of the canal to satisfy the discharge and water elevation targets;
- to test automatic operational procedures and evaluate their efficiency.

The model is built around three main computer programs (TALWEG, FLUVIA and SIRENE) that respectively carry out the topography and geometry generation, the steady and unsteady flow computations. The SIC model for simulation of sediment in irrigation canals needs a sediment inflow rate for each reach, it can be:
- an imposed value;
- a function of sediment load in its upstream reach.

ISIS-Sed is the mobile bed sediment transport module used with ISIS Flow for studying the morphology of rivers and alluvial canals (Halcrow, 2003). The model was developed by HALCROW and HR Wallingford. It has been applied to engineering problems and for studying sedimentation problems in rivers and canals,

including a number of major irrigation systems. The ISIS-Sed software predicts (Halcrow, 2007):
− sediment transport rates;
− uniform or graded sediments;
− changes in bed elevation and channel geometry;
− amounts of erosion and deposition.

Rates of sediment transport are predicted using one of four sediment transport equations, developed for a range of bed material sizes from either empirical studies or from physical considerations. The conveyance properties of the river sections are updated throughout the simulation based on the predictions of the module. ISIS Sediment also includes the ability to simulate dredging of the channel at periodic intervals during a simulation.

HEC RAS model has been developed by US Army Corps of Engineers (USACE, 2006) This model contains four one-dimensional river analysis components for:
− steady flow water surface profile computations;
− unsteady flow simulation;
− movable boundary sediment transport computations;
− water quality analysis.

A key element is that all four components use a common geometric data representation and common geometric and hydraulic computation routines. In addition to the four river analysis components, the system contains several hydraulic design features that can be invoked once the basic water surface profiles are computed.

Delft3D is a three dimensional model developed by Delft Hydraulics. For modelling the transport of cohesive and non-cohesive sediments, a module SED is available. It can be used to study the spreading of dredged materials and sediment/erosion patterns. The SED module can be applied in all geographic regions where Delft3D is used. The module is generally used to calculate the short-term transport of sediment and sand. In particular when the effect of changes in bottom topography on flow conditions can be neglected. For long-term development of the bottom topography or coastal morphology, a separate morphological module with advanced on-line coupling capabilities with the FLOW and WAVE modules should be applied (Delft Hydraulics, 2007).

4
Assessment of parameters for canal design

4.1 Background

The design discharge and sizing of the canal for modern irrigation schemes should be based upon the present concept of crop based irrigation demand, water delivery schedules and water allocation to the tertiary units. The selection of a crop depends upon the soil type, water availability, socio-economic setting and climatic conditions. The type of crop together with the soil type determines the irrigation method and irrigation schedules, while the type of crop and climatic condition determines the irrigation requirement. The water delivery schedules and the flow control methods are then based on the operation mode of the scheme.

Since, the water level for a given discharge is an important parameter in an irrigation canal, the understanding of roughness, the factors that affect the roughness, the appropriate methods to determine and include it in the design is important. A close prediction of the actual roughness during the design will help to reduce the need of frequent gate operation to maintain the design water level. The sediment transport process in the canal will be closer to the design conditions and hence, the improvement in the design.

Most of the sediment transport predictors consider the canal with an infinite width without taking into account the effects of the side walls on the water flow and the sediment transport. The effect of the side wall on the velocity distribution in lateral direction is neglected and therefore the velocity distribution and the sediment transport are considered to be constant in any point of the cross section. Under that assumption a uniformly distributed shear stress on the bed and an identical velocity distribution and sediment transport is considered. The majority of irrigation canals are non-wide and trapezoidal in shape with the exception of small and lined canals that may be rectangular. In a trapezoidal section the water depth changes from point to point in the section and hence the shear stress. The effect would be more pronounced if the bed width to water depth ratio (B-h ratio) is small. An understanding of the effects of the side slopes and B-h ratio on the sediment transport process will be helpful to make necessary improvements in the sediment transport predictors for their use in irrigation canals.

This chapter presents the assessment and analysis of design discharge calculation process, the roughness determination methods and computation of equivalent roughness, the effects of various geometric and sediment characteristics of a canal on the predictability of the predictors, the effect of non-wide irrigation canals on sediment transport prediction and suggested correction to incorporate the effects.

4.2 Irrigation water demand and scheduling

Crop water requirement

The amount of water required to compensate for the evapotranspiration from cropped fields is defined as the crop water requirement (FAO, 1998). The evapotranspiration (ET) is the combination of two processes; evaporation and transpiration. The contribution of each process in the total ET keeps on changing with the growth of the plants. At sowing, nearly 100% ET comes from evaporation, while at full crop cover more than 90% ET comes from transpiration. The amount of ET depends upon climatic conditions and the process is affected by the availability of water (management practices) and vegetation cover (type and stage of crop). The reference crop evapotranspiration (ET_0) is defined as the evapotranspiration rate from a reference surface (grass covered), not short of water. ET_0 values measured or calculated at different locations or in different seasons are comparable as they refer to the ET from the same reference surface. The crop water requirement is then computed as:

$$crop\ water\ requirement = K_c * ET_0 \qquad (4.1)$$

Where, K_c is the crop coefficient and includes the effects due to crop type, variety and development stage. Crop water requirement under standard conditions refers to the evaporating demand from crops that are grown in large fields under optimum soil water, excellent management and environmental conditions, and that achieve full production under the given climatic conditions.

Irrigation method

The requirement of water to the crops can be provided in various ways depending upon the soil type, crop type, farm sizes and topography. Irrigation water can be conveyed, distributed and applied either by a gravity system, a pressurized system or a combination of both methods. In case of surface irrigation, common methods are basin, furrow and border (FAO, 1988). Basin irrigation method is most common for rice growing fields and in the hills. The farmland is divided into small flat parcels surrounded by bunds. Levelling is important as rice requires standing water and an uneven land surface might lead to reduction in production. While, for dry foot crops the movement of water within the farm is slow in case the surface is made flat. Furrow and border irrigation methods are better for all dry foot crops.

The choice of irrigation method is important from a design perspective, since different methods have different achievable application efficiencies. Application efficiency is the fraction of the water volume applied to a farm or a field that is "consumed" by a crop, relative to the amount of water applied at the field inlet (FAO, 1997).

Crop calendar

When the crop water requirements are known, irrigation requirements can be estimated in order to determine the flows that are needed at any moment and place to satisfy the water requirements of the plants. This requires information regarding the types and area of crops grown in the command area and the amount of effective rainfall.

The type and area of crops to be planted in the command area depends upon various factors. Availability of water, climate, soil type, socio-economic setting, country policy and culture determine the type of crops. To work out the flow rate a crop calendar (Figure 4.1) is prepared that shows the type of crops, area to be covered and planting and harvesting dates. The crop calendar bears significant importance for an irrigation scheme, since it reflects the objectives set by the Government or the farmers (farmers association). It is also used to compute the benefit and hence to justify the investment in construction or modernization.

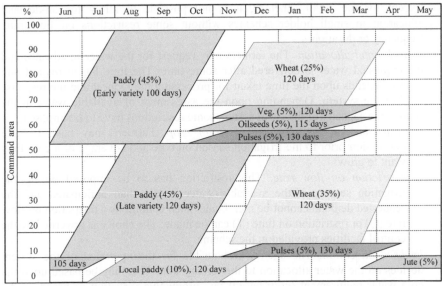

Figure 4.1   Typical cropping calendar for an irrigation scheme in the Terai of Nepal.

Total irrigation requirement

The net irrigation requirement (NIR) is the net crop water requirement after deducting the effective rainfall and taking into account the application efficiency. The total irrigation requirement (TIR) at the inlet of one tertiary unit is:

$$\text{TIR} = \sum_{crops} NIR * \text{area covered} \qquad (4.2)$$

The delivery of water to a field to meet the total irrigation requirement includes three variables; discharge (Q), duration (T_d) and interval (T). Theoretically there are several combinations of Q, T_d and T possible for the supply of water. The selection of any combination depends upon the objectives of the scheme, water availability, cropping policies, etc. and determines the size of the tertiary canal as well as the type of control structures to be provided.

Water delivery schedules

The duration, frequency and amount of water supplied to meet the irrigation requirements is known as irrigation scheduling. The scheduling of irrigation flows may be supply-oriented, demand oriented or a combination of both (Dahmen, 1999). Irrigation scheduling is mostly determined on the basis of a decision making procedure or who decides on the allocation of water to the tertiary unit (Ankum, 2004). There are three possible allocations:

– *on demand allocation*. The farmers have access to the water as and when required. In order to operate the scheme on-demand there must be storage of water at the source or along the canal (in canal storage). Moreover the response time of the canal must be rapid. This system requires at least four times the farm outlet capacity of a continuous flow supply-schedule system (Laycock, 2007). Except for the small schemes, it is not possible to allow for on-demand allocation in small holder schemes without some restrictions on flow rate, frequency or duration;

– *semi-demand allocation*. The farmers may request for the water in advance and the requested water is delivered after some time. The time lag in request and supply depends upon the time taken for processing the request and the response time of the system. Depending upon the assessment of available water in the source, the capacity of the canal system some restrictions have to be imposed:

   o *restriction on the type and area of crop*. Farmers may have to take permission prior to the irrigation season on type and coverage of a crop they want to grow;

   o *restriction on flow rate*. This restriction has to be imposed during the irrigation season, if the available water is less than anticipated and the requested demand cannot be met. Under such conditions a restriction on flow rate (Q) or restriction on time (T) can be made. The choice is governed by the control facilities provided in the scheme.

– *imposed allocation*. The agency responsible for the operation of the system decides on the water allocation to the tertiary unit. This may be based upon the crop water requirement (crop based) i.e., productive irrigation or based on the available water in the source (supply based) i.e., protective irrigation. On crop based allocation, the water delivery plan is determined before the start of the irrigation season and it normally is guided by the designed cropping calendar of the scheme. The farmers have to follow the crop calendar for a smooth operation of the scheme. Since the amount and duration of irrigation gifts is fixed, there is no flexibility on the part of users in the selection of the crops and area to be covered. In supply based allocation, the available water is distributed over the area irrespective of the demand. The time is fixed, while the amount depends upon the available water in the source.

4.3   Estimation of design discharge

The design discharge depends upon various factors and the estimation process can be summarized as shown in Figure 4.2. Soil type, water availability, socio-economic condition of the area and climate are the major factors that separately or in combination determine the irrigation method and irrigation requirement which in turn determine the design discharge. The selection of design discharge and interrelationship of various factors in the derivation of design discharge is more important for the modernization of an irrigation scheme that has been designed and operated differently. Majority of the old irrigation schemes in Nepal were designed for providing the supplementary irrigation with an objective of protective irrigation. Theses schemes have low duty, offered no flexibility in the selection of crops and water demand. During the modernization, the irrigation schemes are being converted into demand based and flexible water delivery mode. Accordingly the canal capacities have been revised and the sections have been designed. In some cases, the modernized systems have been found to be not performing as per expectation mainly due to the:

– lack of enough investigation on the socio-economic dimension of the users to cope with the new operation mode and the irrigation practices prevailing in the area;
– insufficient available water in the main system to meet the proposed water delivery and control plans to the tertiary canals;
– high operation and maintenance cost.

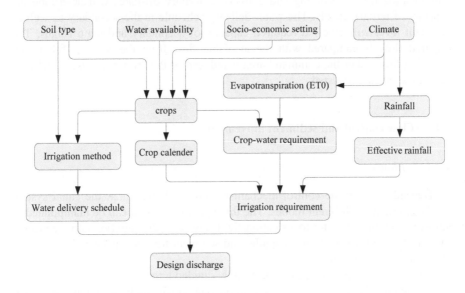

Figure 4.2   Process of the design discharge calculation.

Sizing of a canal

The capacity of a tertiary canal depends on the method of water allocation to a tertiary units. The water allocation method may be:

– *proportional allocation.* In proportional allocation the maximum size of the tertiary canal is the sum of the peak demand for the lower order canals. Proportional division structures are based upon the area to be covered. From management perspective this is the simplest system to manage and least manpower is needed. The size of the canal keeps on decreasing in the downstream direction. All the canals receive water irrespective of the demand;
– *on/off allocation.* The tertiary canals receive water in turns. The supply is either full flow or no flow. The sizing of a canal depends upon how water is distributed in the lower order canals. If all the lower order canals receive water simultaneously, the size of the canal should be sufficient to carry all the cumulative peak discharges. This type of allocation can be used in case of imposed allocation based on crop water demand of the cropping plan. Semi-demand allocation is not possible, since the capacity of the canal is fixed on the basis of maximum peak allocation, and there are no control and measuring facilities in the lower order canal;

− *adjustable allocation.* The amount of flow can be adjusted, depending upon the demand. The lower order canals should be provided with discharge measuring and gated control to regulate and measure the flow. Semi-demand or on-demand water allocation can be made, the limitation, however, is the provision of storage.

The calculated irrigation requirement is used to determine the flow rate depending upon the irrigation method in the field, type of soil and type of crop. The type of allocation, scheduling and delivery at tertiary off-takes will determine the capacity of tertiary canals. The water delivery schedule in the secondary as well as the main level will give the required capacity of secondary and main canals. The required flow is compared with the available flow from the source. If it is not matching then either the command area is reduced or the cropping pattern, delivery schedules, etc. are adjusted.

4.4 Flow control and sediment movement

4.4.1 Flow control

Depending upon the control mechanism and available water, the canal system can be operated in different modes. The three parameters that can be manipulated to adjust the supply of water to irrigation fields are; amount, duration and frequency. The most complex case from hydraulic and sediment transport perspective is when; all the three can be changed.

Irrigation canal operations can be broadly divided into:

− *continuous flow.* The situation is a simple one, in which all the canals are opened simultaneously. The lateral flow may be varied during the operation. Available water is distributed proportionally to all the area;

− *on-off or rotational flow.* Different combinations of rotation are possible. Depending upon the water requirement and available water in the canal the laterals can be grouped in two or more groups and opened and closed simultaneously. It is possible to model each lateral with a separate operation plan, however, this is seldom practiced in reality, simply because of the difficulty in operation and management.

− *adjustable flow.* The adjustment in flow rate may be necessary from the change in demand or change in the available water in the source.

Flow control systems manage the water flows at bifurcations to meet the service criteria and standards regarding flexibility, reliability, equity and adequacy of delivery. Flows can be regulated through water level control, discharge control, and/or volume control. A combination of water level control and discharge control is most common in irrigation systems. Flow rates at off-takes are often indirectly controlled through water level control in the conveyance canals. Variation in discharge at the off-take is determined by the variations in upstream water levels and the sensitivity of the off-takes and water level regulators to those variations.

The methods of control and division of water can be classified according to the orientation of the control (upstream, downstream and volume control), the degree of automation (manual, hydraulic or electrical automatic control) and the form of control (local or central control).

Upstream control

Upstream control is by far the most commonly used type of control around the world. Controlled flows are released to a reach from the upstream end as per the pre-arranged schedule. The water level in the reach is maintained using a water level regulator at the downstream end of the reach. The off-takes are located upstream of the water level regulator. Since, the flow rates are controlled from the upstream end, increment in the off-take demand cannot be entertained unless extra flow is released from the upstream end. This mode of operation forms a negative dynamic storage within the canal reach (Figure 4.3). From operation point of view, this storage is unusable to meet the immediate demand of the off-takes. Moreover, when there is sudden decrease in demand and off-take gates are closed, this stored water has to be released either to the drain or towards downstream reaches that may or may not be utilized. Upstream control is a serial control, is supply oriented, has limited flexibility and requires well equipped centralized management setup for running the system properly.

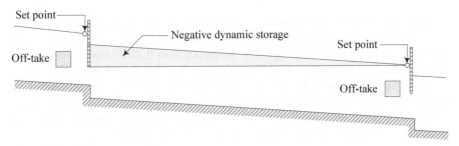

Figure 4.3   Upstream control.

Upstream control can be made fully automatic like predictive control or Electronic Filter Level Offset (ELFLO) control (Ankum, 2004). This control provides an option for semi-demand allocation to tertiary off-takes. The automation is suitable when there is sufficient water in the source and reliable power supply for the operation of electro-mechanical gates. The major advantage of automation is that it is self managed and reduces the response time of the system as management delays in responding the change in demand are minimized. However, managing automatic predictive control in a long canal with series of controller is quite complex and should be checked using hydro-dynamic flow models (Schuurmans, *et al.*, 1999).

Downstream control

Downstream control systems respond to water level changes downstream of a regulator. They are designed to permit instantaneous response to changes in demand by using water in storage in the upstream canal section (positive dynamic storage) in the case of an increase in demand and by storage of water in case of decreasing demand. Downstream control systems can be manually operated, but they are easily automated, either hydraulically (Neyrpic) or electrically. AVIS and AVIO (commercial names) are the examples of automatic gates that operate on the water level in the downstream side (Ankum, 2004).

Both water level and flow are controlled at the upstream end of a canal section (Figure 4.4). Changes are gradually passed on in upstream direction till the head works. Most downstream control systems use balanced gates. Downstream control systems are specifically designed to maximize flexibility by minimizing system response times. The hydraulic stability of the system should be checked, preferably by a mathematical model.

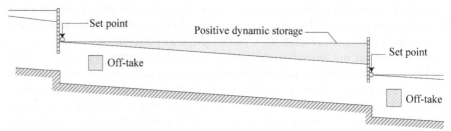

Figure 4.4   Downstream control.

A considerable raising of berms is needed in steep terrain; this can make the system very expensive. The application of downstream control is, therefore, usually limited to canals with a bed slope that is smaller than 0.3‰. Downstream control cannot be applied in canals with drops (structures that cater for drops in water level that are greater than the hydraulic losses over the structure). Downstream control can be applied both in on-demand and on-request delivery systems. The main design requirement is that each canal section has sufficient capacity to meet the maximum instantaneous demand. Downstream control is also a serial process.

Volume control

The volume control method also known as BIVAL control involves the (usually simultaneous) operation of all flow control structures in the system to maintain a nearly constant volume of water in each canal section (pool) (Ankum, 1993). This method is used to meet operational requirements for different users of water in a flexible manner as it can provide immediate response to changing demand. With this type of operation, the water surface between control structures rotates around a point located approximately midway between the control structures that are operated to respond to changes of demand in the canal section (Figure 4.5). Main advantage of this control is in the reduction of wedge storage (Laycock, 2007). It works on the premise that at zero discharge the water surface will be horizontal and at maximum discharge the water surface will be parallel to the bed, but pivot around the mid point of the canal reach. Implementing this type of control requires automation of the control structures coupled with centralized monitoring and control. With remote control all control points in the system can be operated both simultaneously and serially.

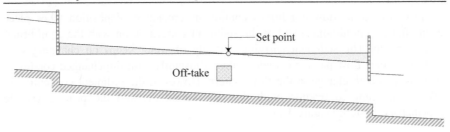

Figure 4.5   Volume control.

4.4.2   Sediment movement

The operation of gates to control and distribute water creates non-uniform flow in an irrigation canal network. In non-uniform flows the sediment carrying capacity along a canal keeps on changing.

A non-equilibrium sediment transport may occur under both uniform and non-uniform flow conditions. Under uniform flow condition (Figure 4.6), when the inflowing water carries less than the equilibrium sediment load then the flow try to achieve the equilibrium by picking up sediment from the bed (case-II(a)). After some distance the readjustment of the sediment concentration takes place and the flow acquires an equilibrium condition. This distance is called the adaptation length. Initially sediment is picked up immediately after the flow enters the erodible region. But, this cannot continue for long as after some time the slope near the head reach becomes more flat and less and less sediment is picked up. Instead the sediment pickup starts from some point downstream and the process is continued until the whole reach of the canal attains equilibrium with the inflowing sediment load. However, if the flow is restricted to pickup sediment from the bed (case-II(b)), then the flow will continue under non-equilibrium conditions.

Similarly, if the inflowing water carries more than the equilibrium sediment load (case-I) then the extra amount is deposited in the canal so that a new equilibrium in the flowing water is attained.

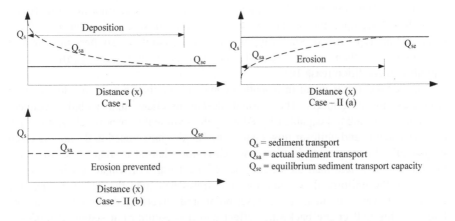

Figure 4.6   Suspended sediment transport pattern under equilibrium condition.

In non-uniform flow conditions continuous erosion or deposition takes place even after an initial adjustment of the sediment concentration with the equilibrium concentration. The sediment carrying capacity of water keeps on changing with distance. Since flow parameters adopt almost instantly with the changed condition, there is a smooth change in the flow pattern. However, the sediment concentration takes some time to readjust with the new conditions and the process can be schematised in steps (Figure 4.7).

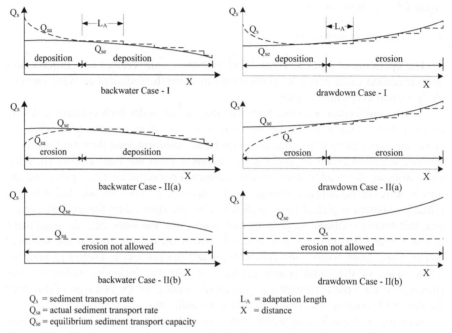

Q_s = sediment transport rate
Q_{sa} = actual sediment transport rate
Q_{se} = equilibrium sediment transport capacity

L_A = adaptation length
X = distance

Figure 4.7   Sediment transport pattern under backwater and drawdown conditions.

In case of backwater condition, there may be continuous deposition (case-I), first erosion and then continuous deposition (case-II(a)) or non-equilibrium flow (case-II(b)) depending upon the actual concentration in the flow, carrying capacity of the canal and the type of bed. Similarly, in drawdown condition also there may be first deposition and then erosion (case-I), continuous erosion (case-II(a)) or non-equilibrium condition (case-II(b).

The sediment movement in irrigation canals is not simple once the flow control structures are introduced. The normal design practices that include sediment transport aspect are not capable of predicting the sediment transport behaviour. They assume a steady and uniform flow condition. The discharge used to design the canal is generally the maximum discharge, which is not released in majority of the time. Once the flow is less than the design the sediment transport capacity will decrease even with the uniform flow condition. For flows less than the design values, the gates are lowered to maintain the set-point and diverting desired water to the laterals. This will create backwater effect and non-equilibrium sediment transport conditions.

Hence the design approach should be modified in such a way that the actual sediment movement scenario is reflected. Then only the designed canal will be able to transport the expected sediment load under changing flow conditions.

4.5   Roughness prediction

4.5.1   General

Roughness relates the mean velocity to the depth of flow, slope and sediment and fluid characteristics. Selection or prediction of a roughness factor for the design of an irrigation canal is crucial, as it also determines the type of maintenance level expected in the future. Hydrodynamic performance of the scheme depends more on how close the predicted and actual roughness are. If the selected roughness for the design is low, then for most of the time the actual roughness will be more than the design value and discharge capacity of the canal will be less than the design. Too frequent maintenance of highest level is needed to meet the design roughness. This requires more maintenance cost and skilled manpower. Designers prefer to choose a lower roughness and smaller canal sections, so that the construction costs are less, since economic indicators are one of the major yardsticks in the selection of an irrigation scheme for investment. On the other hand, if the design roughness is more rough than the actual one, the normal depth for the design discharge will be less. A larger number of water level regulators is needed to raise the water level so that the targeted command is irrigated. This not only increases operation cost of the scheme, but also changes the flow pattern and the sediment transport behaviour of the canal.

Roughness of a canal section can be measured under uniform flow conditions, which can be the basis for the design of similar other canals or a calibration constant for modelling. Measured values are overall roughness of the whole section due to the influence of various factors. Hence it is important to know how each factor influences the total effective roughness of a canal. Major factors that influence the roughness of a canal are (Chow, 1983):

−   size and shape of the material forming the wetted perimeter;
−   vegetation in the bed and sides;
−   canal surface irregularities. These irregularities may be due to localized erosion/deposition, bed forms, poor canal construction or maintenance, etc;
−   shape of canal and bed width to water depth ratio;
−   degree of variation in canal alignment, size, shape and cross section in the longitudinal direction.

Depending upon whether the canal has a rigid boundary or erodible boundary, carries clear water or water with sediment, the influence of various factors will be different. Irrigation canals are man-made, so the degree of variation in canal alignment, size, shape and cross section are normally low. Hence the effect of variation in alignment, size and shape along the canal can be ignored. For the analysis of a canal carrying sediment load with erodible boundary, the bed can be assumed to be free from vegetation. Hence the effect of vegetation will be in the sides only. The irregularities in the surface may be both in the bed and sides, but the source of irregularities may be different. If the irregularity is due to the bed forms then it will be in the bed only, while irregularity due to poor construction or maintenance will be mostly in the sides. Once the canal is operated the bed will be smoothened and irregularities will be bed forms. Hence, the roughness is not uniform throughout the perimeter except for the ideal canal. If the roughness is not the same over the perimeter then the shape of the canal and bed width to water depth ratio (B-h ratio) will also influence the overall roughness of the section. Figure 4.8 shows the effective overall roughness computation process in a erodible boundary canal.

It should, however, be realized that the influence is not constant for different flow conditions and with time. Bed forms are the function of flow and sediment characteristics. The condition of a canal does not remain the same during operation and vegetation also keeps on changing. Moreover, there might be periodic maintenance plans during the operation. Hence a methodology should be employed to predict the roughness with time to study how the designed canal will perform during its operation.

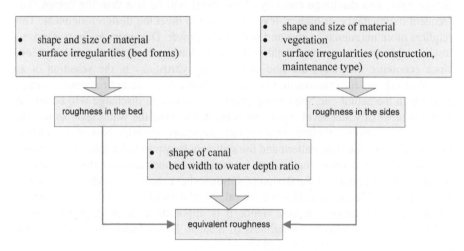

Figure 4.8   Process to derive the equivalent roughness.

4.5.2   Roughness in the bed

In alluvial canals there exists two stages of flow, one when there is no movement of bed material and the other when bed material is moving. No movement of bed material can be compared with the condition of a rigid boundary canal having an equivalent roughness height (k_s) equal to the representative bed material size (d). The resistance to flow in a movable bed consisting of sediment is mainly due to grain roughness and form roughness. Grain roughness is generated by a skin friction force and form roughness by a pressure force acting on the bed forms. Since the bed forms change continuously with the flow parameters (velocity, depth), the bed roughness also changes (Figure 4.9). There are two approaches to estimate the bed roughness:

– methods based on hydraulic parameters such as mean depth, mean velocity and bed material size;

– methods based on bed form and grain-related parameters such as bed form length, height, steepness and bed-material size.

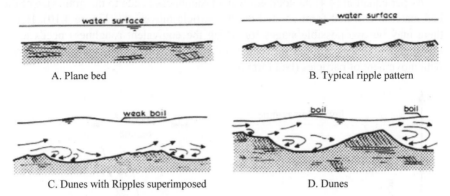

A. Plane bed   B. Typical ripple pattern

C. Dunes with Ripples superimposed   D. Dunes

Figure 4.9   Forms of bed roughness in mobile bed in lower flow regime (Simons, *et al.*, 1992).

Methods based on hydraulic parameters

The methods proposed by Einstein and Barbarossa (1953), Engelund (1966), White, *et al.* (1980), Brownlie (1981) are most widely used ones. These methods do not take into account the shape and size of bed forms explicitly in the prediction of the roughness. Brownlie (1981) relates discharge, slope and sediment characteristics with depth of flow. The known flow depth then can be used to determine the roughness of the section. White, *et al.* (1980) give a relation to compute the mean velocity from water depth, slope and sediment characteristics, which can then be used to predict the friction factor. Engelund (1966) gives a relation to compute mean flow velocity with slope, water depth, mobility parameter.

Methods based on bed form parameters

Methods based on bed form parameters separate the total roughness into that due to grains and bed forms. These methods require the shape and size of bed forms and this information is used explicitly in the determination of the equivalent roughness height. Mendez (1998) tested the accuracy of the friction factor predictors with selected data and concludes that the method by Van Rijn (1984b) based on bed form parameters gives the best results.

The total effective shear stress (τ) can be separated in two parts (Figure 4.10), namely that due to skin (τ') and form (τ'') resistance:

$$\tau = \tau' + \tau'' \tag{4.3}$$

Similar to the separation of shear stress as grain and form, Nikuradse's equivalent sand roughness height (k_{se} in m) can be separated as (Van Rijn, 1982):

$$k_{se} = k_s' + k_s'' \tag{4.4}$$

where

k_s' = roughness due to grain (m)

k_s'' = roughness due to form developed in the bed (m)

As per equation (4.4) the total equivalent roughness is due to the grain size of the bed material and bed form created due to particle movement (Figure 4.10). Hence there may be two possible stages for which the equivalent roughness needs to be investigated. One is without movement of particles (plane bed) and the other one with movement of particles (bed forms).

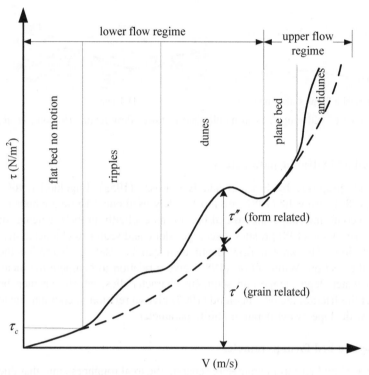

Figure 4.10 Schematization of the total shear stress due to grain and bed form as a function of mean velocity (Van Rijn, 1993).

Roughness due to grain can be determined as some multiple of representative bed material size. However, there is no general agreement on which size of the bed material should be taken as representative size. The recommended values range from $k_s = 1.25d_{35}$ (Ackers, et al., 1973), $k_s = 5.1d_{84}$ (Mahmood, 1971) to $k_s = 2.5d_{90}$ (Kamphuis, 1974). Van Rijn (1982) proposes the roughness due to grain as:

$$k_s' = 3 * d_{90} \tag{4.5}$$

Assuming a regular sediment size distribution in the irrigation canal:

$$d_{90} = 1.5d_{50} \tag{4.6}$$

then,

$$k_s' = 4.5d_{50} \tag{4.7}$$

When the velocity increases the material starts moving and as a result the bed feature changes. In irrigation canals the minimum velocity is limited to 0.10 to 0.15 m/s (Dahmen, 1994), below which the canal becomes unnecessarily wide and uneconomical. Similarly the maximum velocity is limited to a value that does not cause undesirable erosion of the bed. Moreover, higher velocity is restricted since it requires a steeper slope that causes loss of elevation which is not desirable. Critical shear stress is generally limited to 3-5 N/m^2. Considering the range of adopted shear stress between 1 to 5 N/m^2 and using Shields' criteria for the initiation of motion, the bed material in irrigation canals is almost always in motion (Figure 4.11). Hence some bed forms can be expected. This change in bed feature consequently changes the roughness. The flow in irrigation canal lies within the lower flow regime with Froude number mostly less than 0.5, hence the bed forms is either ripples, mega ripples or dunes (Figure 4.10).

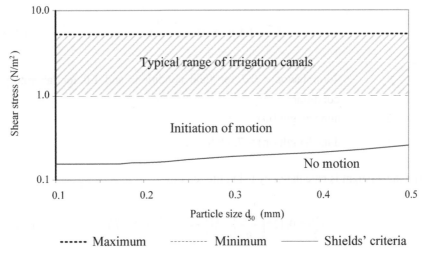

Figure 4.11   Comparison of initiation of motion (Shields' criteria) and shear stress range in irrigation canals.

According to Van Rijn the second term in equation (4.4) is related to the bed form height (Δ), the bed form steepness (Δ/λ) and the bed form shape (γ) (Figure 4.12). The following functional relationship is assumed to be valid:

$$k_s'' = f(\Delta, \Delta/\lambda, \gamma) \tag{4.8}$$

for ripples
$$k_s'' = 20\gamma_r\Delta_r\left(\frac{\Delta_r}{\lambda_r}\right) \tag{4.9}$$

where

Δ_r = ripple height = 50 to 200 * d_{50} (m)

γ_r = ripple presence ($\gamma_r = 1$ for ripples only)

λ_r = ripple length = 500 to 1000 d_{50} (m)

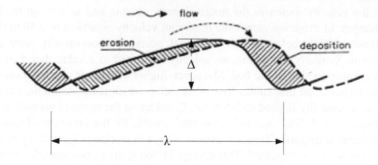

Figure 4.12   Typical bed form and migration pattern (Van Rijn, 1993).

Similarly, for dunes:

$$k_s'' = 1.1 \, \gamma_d \, \Delta_d (1 - e^{-\frac{25\Delta_d}{\lambda_d}})$$ (4.10)

where

γ_d = form factor = 0.7 for field conditions and 1.0 for laboratory conditions

Δ_d = dune height (m)

λ_d = dune length (m) = 7.3 * h

The dune height is calculated by using the relation:

$$\frac{\Delta_d}{h} = 0.11 * \left(\frac{d_{50}}{h}\right)^{0.3} * \left(1 - e^{-0.5*T}\right) * (25 - T)$$ (4.11)

where, T is the excess bed shear parameter given as:

$$T = \frac{u_*^2 - u_{*cr}^2}{u_{*cr}^2}$$ (4.12)

$$u_* = \frac{V\sqrt{g}}{C'}$$ (4.13)

$$u_{*,cr} = \sqrt{\frac{\tau_{cr}}{R}}$$ (4.14)

where

C' = Chézy's roughness coefficient related to grain (m$^{1/2}$/s)

R   = hydraulic mean radius (m)

T_{cr} = critical shear stress (N/m^2)

V   = flow velocity (m/s)

For the determination of the critical shear velocity (u_{*cr}) the Shields' diagram (Figure 3.2) as well as the condition for the initiation of motion as proposed by Van Rijn can be used. The dimensionless particle parameter (D_*) and excess bed shear parameter (T) is related with the type of bed forms (Table 4.1).

Table 4.1   Classification criteria of bed forms (Van Rijn, 1984c).

Bed forms	D*	T
Ripples	$1 \leq D^* \leq 10$	$0 \leq T \leq 3$
Mega-ripples	$1 \leq D^* \leq 10$	$3 < T \leq 10$
Dunes	$D^* > 10$	$T > 10$

Once the initiation of motion starts the ripple will be formed and with ripple height as 100 * d_{50} and ripple length as 1000 * d_{50}, gives k_s = 200 * d_{50}. Taking the shear velocity (u_*) as the critical mobility shear velocity from Shields' criteria, the flow regime of an irrigation canal for sediment in the range from 0.05 mm to 0.50 mm will be rough once the initiation of motion takes place (Figure 4.13).

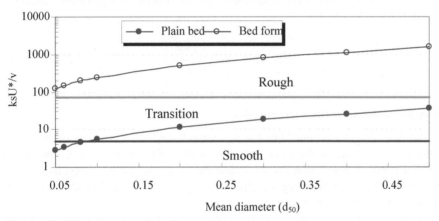

Figure 4.13   Flow regimes with sediment size.

Then the Chézy's roughness coefficient for different flow regimes is given by equations (3.30 to 3.32). The computed roughness height can be used in the corresponding equations (transition or rough regime) to determine the Chézy's roughness coefficient.

4.5.3   Roughness in the sides

Roughness in the side slopes of a canal depends upon the size and shape of the surface material, irregularities in the surface and vegetation.

Roughness due to surface material

The equivalent roughness height due to the material in the surface of a canal section is related to the median particle diameter (d_{50}) as 4.5 * d_{50}. This value is for a rigid boundary canal. However, a movable bed of loose material will have a somewhat larger roughness because a perfectly plane bed will not exist in natural conditions and small irregularities will always be present. This was also shown by Lyn (1991) in his experimental results. Hence Van Rijn (1993) suggests to take a minimum value of k_s as 0.01 m. Moreover the Manning's roughness coefficient for an ideally finished earthen canal is 0.016 to 0.020 (Chow, 1983). Manning's roughness coefficient (n) is an average value and is assumed to be uniform over the entire cross section. The value of n can be related to Chézy's roughness coefficient for the side part only by the relation:

$$C_s = \frac{R_s^{\frac{1}{6}}}{n} \qquad (4.15)$$

where

C_s = Chézy's roughness coefficient in the sides only ($m^{1/2}/s$)
R_s = hydraulic mean radius in the sides (m)

Now, Chézy's roughness coefficient can be related to equivalent roughness height in the sides (k_{ss}) as:

$$C_s = 18 \log\left(\frac{12 R_s}{k_{ss}}\right) \qquad (4.16)$$

Roughness due to surface irregularity

Surface irregularity may be due to methods and workmanship of the canal construction, ageing of the canal, rain-cuts, slides of the banks, etc. Chow (1983) gives a detailed description on the classification of the surface irregularities and correction factors for Manning's coefficient. As per his classification the surface can be divided into four categories:

– *ideal*. This refers to the best attainable surface for the construction material. Newly constructed or maintained canals with perfect workmanship can be covered under ideal type of canal. No correction is needed for such a type of surface. For earthen canals the value of Manning's n is 0.018;
– *good*. In this category fits newly constructed or weathered but well maintained canals with good to moderate finishing. A value of 0.005 is added for this surface that makes the value of Manning's n as 0.023;
– *fair*. The surface of the canals that are moderate to poorly excavated are under this category. It also includes the canals that have been excavated by machines and have eroded side slopes. A value of 0.01 is added for this surface (n = 0.028);
– *poor*. Badly eroded or sloughed side slopes, large rain cuts and excavation not in proper shape. Add 0.02 to Manning's n for this surface (n = 0.038).

Roughness due to vegetation

Vegetation reduces the effective flow area and increases the roughness of a canal section. Growth of vegetation is more pronounced in clear water, however the nutrients with the sediment may be helping to grow weeds even in sediment laden flows. The degree of obstruction created by vegetation is highly variable and depends upon the type, height, density and flexibility of vegetation, submerged or un-submerged condition, water level, flow velocity (Petryk and Bosmajian, 1975, Kouwen and Li, 1980, Kouwen, 1988, Querner, 1997, Wu, *et al.*, 1999). Kouwen (1988) gives a simple empirical relation to calculate the equivalent roughness height for a given vegetation:

$$k_s = 0.14 h_g \left[\frac{(mei / \tau)^{0.25}}{h_g} \right]^{1.59} \tag{4.17}$$

mei is a parameter that is a function of stem density, modulus of elasticity and second moment of area and is given by:

for green grass: $\qquad\qquad mei = 319 h_g^{3.3}$ $\qquad\qquad$ (4.18)

for dead grass: $\qquad\qquad mei = 25.4 h_g^{2.26}$ $\qquad\qquad$ (4.19)

where

$\qquad h_g$ $\qquad$ = local height of the vegetation (m)

$\qquad k_s$ $\qquad$ = equivalent roughness height (m)

$\qquad \tau$ $\qquad$ = local boundary shear stress (N/m^2)

Validation of the coefficients using measured field data is needed to use equation (4.17) for conditions other than that for which the equation was derived.

Table 4.2   Weed factor for different types of vegetation at full growth (derived from Chow's (1983) suggested n for vegetation).

Category	Description	Weed factor
Low	– dense growth of flexible turf grass (h/h$_g$ = 2-3) – supple seedling tree switches (h/h$_g$ = 3-4)	1.25 - 1.5
Medium	– turf grasses (h/h$_g$ = 1-2) – stemmy grasses, weeds or tree seedling (h/h$_g$ = 2-3) – brushy growths, moderately dense	1.5 - 2.5
High	– turf grasses (h/h$_g$ = 1) – willow or cottonwood trees 8-10 years old; – bushy willows	2.5 - 3.5
Very high	– turf grass (h/h$_g$ = 0.5)	3.5 - 6.0

Note: h = water depth and h$_g$ = height of vegetation.

Another simple way of incorporating the effect of weed is to follow the suggestion by Chow (1983). He has given the classification of vegetation and the value of correction to be added to the Manning's roughness coefficient. Table 4.2 summarizes the classification types, description of the stage and type of vegetation of each category and corresponding factor (weed factor) based on his criteria.

4.5.4   Equivalent roughness

Irrigation canals are normally not very wide (B-h ratio < 8), hence the influence of the roughness in the side slope will be significant in the overall roughness value. Different methods are available to determine equivalent roughness (k_{se}) of a section. Yen (2002) discussed in detail the different aspects of computing composite roughness of an open canal.

Method 1. In this method the mean velocity and the energy gradient at each subsection is assumed to be the same. It was proposed independently by Horton and by Einstein (Chow, 1983).

$$n_e = \left(\sum_{i=1}^{N} \frac{P_i n_i^{\frac{3}{2}}}{P} \right)^{\frac{2}{3}}$$

(4.20)

where

n_e = the composite n value for the whole section
n_i = n value for the subsection
N = total number of subsections
P_i = wetted perimeter for the subsection (m)
P = total wetted perimeter in the cross section (m)

Method 2. This method was proposed by Pavlovskiĭ, by Muhlhofer, and by Einstein and Banks (Chow, 1983). It is based on the hypothesis that the total force resisting the flow is equal to the sum of the forces resisting the flow in each subsection. The resulting composite roughness value is:

$$n_e = \left(\sum_{i=1}^{N} \frac{P_i n_i^2}{P} \right)^{\frac{1}{2}}$$

(4.21)

Method 3. Ida (1960) derived a relation by equating the discharge through all the subsections with the whole section. He used the composite R in place of the mean R and the composite roughness value is given by:

$$n_e = \frac{\sum\limits_{i=1}^{N} P_i R_i^{\frac{5}{3}}}{\sum\limits_{i=1}^{N} \frac{P_i R_i^{\frac{5}{3}}}{n_i}}$$

(4.22)

Method 4. Krishnamurthy and Christensen (1972) proposed that the summation of the discharges in subsections with roughness coefficient $k_{s,i}$ (i subscript for the subsection) is equal to the summation of the discharges of all the subsections with an equivalent composite roughness k_{se}. The flow in each section is assumed to be rough turbulent and the velocity distribution is described by the logarithmic law. The equivalent roughness value is given by:

$$\ln n_e = \frac{\sum_{i=1}^{N} P_i R_i^{\frac{3}{2}} \ln n_i}{\sum_{i=1}^{N} P_i R_i^{\frac{3}{2}}} \qquad (4.23)$$

Method 5. Méndez (1998) proposed to divide the discharge into the discharges through the central part and side slope parts, such that the total discharge is the sum of subsection discharges.

$$Q = Q_{cen} + Q_{lat} \qquad (4.24)$$

Neglecting the effect of the momentum transfer, the discharge through a stream column of width dy and water depth h_i, with a local roughness height k_{si} can be calculated as (Figure 4.14):

$$Q_i = C_i h_i \, dy \sqrt{h_i S_f} \qquad (4.25)$$

where

$$C_i = 18 log \frac{12 h_i}{k_{si}} \qquad (4.26)$$

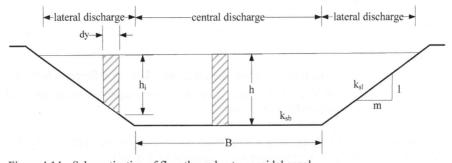

Figure 4.14   Schematization of flow through a trapezoidal canal.

The discharge through each of the subsections is the summation of the flow in each stream tube and can be written as:

$$Q = Q_{lat} + Q_{cen} = 2\int_0^{mh} C_i h_i \sqrt{h_i S_f}\, dy + \int_0^{B} C_i h \sqrt{h S_f}\, dy \qquad (4.27)$$

Solving the equation gives:

$$Q = \frac{18 S_f^{\frac{1}{2}} h^{\frac{3}{2}}}{2.3}\left[\frac{4}{5} mh\left(\ln\frac{12h}{k_{sl}} - \frac{2}{5}\right) + B\ln\frac{12h}{k_{sb}}\right] \qquad (4.28)$$

If the roughness height in the bed (k_{sb}) and sides (k_{sl}) is the same and is equal to k_{se} then equation (4.28) can be written as:

$$Q = \frac{18 S_f^{\frac{1}{2}} h^{\frac{3}{2}}}{2.3}\left[\frac{4}{5} mh\left(\ln\frac{12h}{k_{se}} - \frac{2}{5}\right) + B\ln\frac{12h}{k_{se}}\right] \qquad (4.29)$$

Comparing equations (4.28) and (4.29) gives:

$$\ln k_{se} = \frac{0.8\ln k_{sl} + \dfrac{B}{h}\ln k_{sb}}{0.8m + \dfrac{B}{h}} \qquad (4.30)$$

$$C_e = 18\log\left(\frac{12R}{k_{se}}\right) \qquad (4.31)$$

Since the lateral transfer of momentum and its effect on the velocity distribution across the canal is not taken into account, equation (4.27) will over predict the flow rate. If this equation is to be used for discharge calculation a correction factor for the velocity distribution has to be applied. The modified effective Chézy's roughness coefficient is then given by:

$$C_e' = f_e C_e \qquad (4.32)$$

where

f_e = correction factor for the effective Chézy coefficient which is a function of B-h ratio, side slope and roughness in the bed and sides (Méndez, 1998)

Method 6. Instead of applying a correction on the basis of velocity distribution the equivalent roughness is computed using equation (4.30) only.

All the 6 methods of roughness prediction were compared using the Krüger (1988) data set. Considering the general conditions of irrigation canals, the following criteria have been used for selecting the data:
- trapezoidal cross section;
- Froude number less than 0.5;
- bed width to water depth ratio less than 8;
- the ratio of equivalent roughness height in the bed and the side wall (k_{sb}/k_{sl}) in the range of 50 to 0.02.

A total of 19 records were selected from the compilation of Krüger. Table 4.3 shows the summary of the selected data.

Table 4.3   Characteristics of selected data.

Test	B-h ratio	Side slope	k_{sl} (mm)	k_{sb} (mm)	k_{sb}/k_{sl}	No of records
1	3.0 - 5.8	1	0.054	1.047	19.800	3
2	1.8 - 5.7	1	8.400	1.047	0.125	8
3	3.9 - 7.9	2	0.054	1.047	19.800	8

Calculation process used in the comparison of the selected data:
- since, it is not possible to measure Manning's roughness coefficient separately for bed and side slope, the representative size of particles (d_{50} or d_{90}) was used to determine the equivalent roughness height (k_s). Different authors have given different estimation methods. Henderson (1966) suggests 2~3 * d_{50}, Van Rijn (1982) suggests 1~3 * d_{90}, while Krüger (1988) has suggested it to be d_{90} for plane bed conditions. For this analysis k_s has been taken equal to d_{90};
- the local Manning's roughness coefficients (n_b, n_l) is determined using the relation (Henderson, 1966):

$$n = 0.031(3.28k_s)^{\frac{1}{6}}$$ (4.33)

- the flow can be divided as hydraulically smooth, transition or rough depending upon the following conditions (Van Rijn, 1993):

smooth flow: $\dfrac{u_* k_s}{v} \leq 5$ (4.34)

transition flow: $5 < \dfrac{u_* k_s}{v} < 70$ (4.35)

rough flow: $\dfrac{u_* k_s}{v} \geq 70$ (4.36)

- as per above given criteria, the flow condition for the selected data set is transition, hence the velocity distribution is affected by viscosity as well as by the bed/side roughness. The Chézy's roughness coefficient for transition flow is given by:

$$C = 18\log\left(\frac{12R}{k_s + 3.3\nu/u_*}\right) \tag{4.37}$$

— average value of the Manning's roughness coefficient is then computed from the k_s value from equation (4.37) and compared with the derived n using different methods.

The comparison between the 6 methods is made on the basis of number of well predicted values within an error band. If K is the error factor then the error band is the range between *measured value/K* and *measured value * K*. If a predicted value lies within the above band then it is a well predicted one. The accuracy of a method is given by:

$$\text{accuracy}(\%) = \frac{\text{number of well predicted values}}{\text{number of total values}} * 100 \tag{4.38}$$

The accuracy of predictability of different methods, for different error factors is given in

Figure 4.15. The result shows that method 6 best predicts the equivalent roughness in a trapezoidal canal section. Hence, it is proposed to use the method 6 for the evaluation of equilibrium roughness of a canal section.

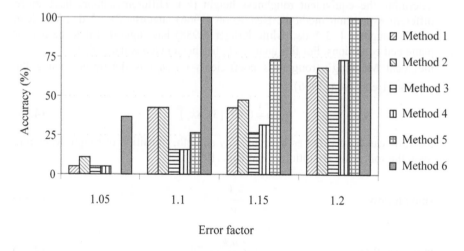

Figure 4.15 Comparison of well predicted values for different values of error factor using Krüger data.

General practice of estimation of roughness and possible errors

Lined canals are designed with a roughness factor specified for the material of lining; for example the roughness coefficient for brick lining is much smaller than that of an earthen canal. The roughness of canals with sediment deposition becomes even higher due to the presence of bed forms. Hence, the discharge capacity of the canal reduces drastically if the sediment is deposited in the lined canal bed, not only by reduction in available depth but also by the increment in roughness (Figure 4.16).

Hence, while designing a lined canal section care should be taken to avoid sediment deposition, for example, by providing steeper canal slope.

Earthen canals are designed with a certain roughness value; when the canal banks have to be protected with boulder pitching or other methods the roughness of the canal section changes and the flow profile is changed. It is not possible to foresee and locate where the protection will be needed during the design. The designer, however, can provide guidelines about what type of protection should be used so that the hydraulic flow in the canal is not affected much. Sometimes, the earthen canals are lined (concrete, bricks, etc.) to minimize seepage losses. Farmers also prefer a lined canal, as they think that the lined canal require less maintenance. Linings reduce roughness and increase conveyance, but at the same time reduce available head. So obstruction has to be placed to raise the head, that is, the gain made in flow velocity by reducing roughness is lost by putting obstruction. Hence, there will be less improvement in terms of sediment transport capacity. Lining will be helpful to prevent erosion problems.

Figure 4.16   Rectangular lined canal with bed deposition.

4.6   Sediment transport in non-wide canals

In general the reliability of sediment transport predictors is low and at best they can provide only estimates. A probable error in the range of 50-100% can be expected even under the most favourable circumstances (Vanoni, 1975). The error is expected to increase further if the calculations are based upon average values of flow and sediment parameters. Several assessments of sediment transport formulas have been made (Brownlie, 1981), (Yang and Molinas, 1982), (Van Rijn, 1984b), (Yang and Wan, 1991) and each provide different results. Woo and Yu (2001) compared the results assessed by different researchers and found that there is no universally accepted formula for the prediction of sediment transport. Most of them are based upon laboratory data of limited sediment and water flow range. Hence they should be adjusted to make them compatible to the specific purposes, otherwise the predicted result will be unrealistic.

The distribution of shear stress along the boundary is not constant as opposed to the general assumptions made in the calculation. Even in laboratory flumes for sediment transport experiments the boundary shear is not constant due to the presence of bed forms. For laboratory data a side wall correction is applied, so that the data can be treated as that of a wide canal and accordingly empirical relationships are developed. Different techniques are in use for the sidewall correction (Einstein, 1942, Vanoni and Brooks, 1957), whose main objective is to find an average shear stress in the bed after making due allowance for the friction of the sidewall.

Engelund and Hansen (1967) have pointed out that even after the theoretical side wall correction, the experimental flume data should be considered with caution. It is because the shear is dependent not only on the relative roughness of bed and walls but also on the width to depth ratio of the flume. Moreover Brownlie (1981) mentions that field data have slightly higher sediment concentrations than laboratory data for similar ranges of dimensional groups. For a theoretical analysis of the discrepancy in laboratory and field observations, he used a typical river section and showed that the difference was due to the changing water depth along the perimeter in the river section compared to a constant depth in laboratory flume.

While analyzing the applicability of the sediment transport predictors in irrigation canals, two aspects should be considered; the side slope and the bed width to water depth ratio (B-h ratio). The majority of the canals has a trapezoidal shape with side slope ranging from 1:1 to 1:4 or even more depending upon the soil type and bank stability with the exception of small and lined canals that may be rectangular. The changing water depth on the sides will have influence in the overall shear distribution along the perimeter. This effect is more pronounced if the B-h ratio is small. Irrigation canals are non-wide in nature, in the majority of the cases the ratio of bed width to water depth is less than 8 (Dahmen, 1994). Hence, the assumption of a uniform velocity and sediment transport across the section and expressing them per unit width of canal does not hold true.

In the following sections the influence of the canal shape on the sediment transport capacity using total load predictors will be investigated. It should be noted that for the same hydraulic and sediment characteristics different predictors give widely varied results. For one specific condition one predictor may be better suited than the other and it is not possible to adjust all the predictors to produce the same value for a given condition. Hence, the purpose of the adjustment is to adapt the equation for a specific canal condition and with that the predictability should be improved.

4.6.1   Effects of canal geometry and flow characteristics in sediment transport

For the flow conditions and sediment characteristics prevailing in the irrigation canals, Méndez (1998), after evaluating the available total load predictors with field and laboratory data, concludes that the predictors given by Brownlie, Ackers and White and Engelund and Hansen are better as compared to other predictors. However, the prediction within an error factor less than 2 was not possible. In this research, mostly the same three predictors will be used for predicting sediment transport under equilibrium conditions.

The sensitivity of the predictors with the flow and sediment parameters has been evaluated. This will provide an insight on how the change in one or more of the parameters will influence the sediment transport capacity of a canal.

Effect of sediment size

The sediment size normally found in irrigation canal is 0.05 mm (50 µm) to 0.5 mm. For the evaluation of effect of sediment size on the predictability of the predictors the discharge, bed width, bed slope, side slope and the roughness of a canal was kept constant and equilibrium concentration was computed for each of the three total load predictors. The predicted value has been presented in Figure 4.17. The Brownlie predictor has almost a linear variation for sediment size of 0.10 to 0.50 mm. Ackers-White gives high values for sediment sizes below 0.15 mm. The predictability of all the three predictors for sediment sizes greater than 0.15 mm is comparable.

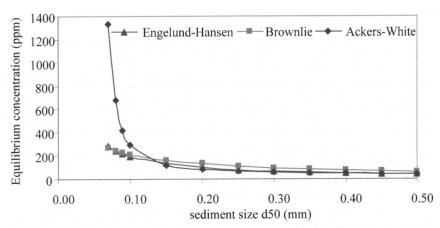

Figure 4.17   Effect of sediment size (d_{50}) on the prediction of the equilibrium concentration.

Effect of velocity

For the evaluation of the sediment transport prediction with the change in velocity, the discharge, bed width and side slope in a canal were kept constant while the slope was increased to increase the velocity. The Froude number was kept below 0.5. The results have been presented in Figure 4.18.

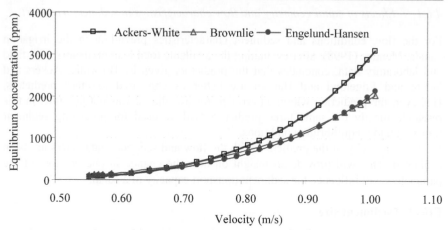

Figure 4.18   Effect of velocity in the prediction of the equilibrium concentration.

The influence of the velocity in the prediction of the transport capacity is almost similar in case of the Brownlie and Engelund-Hansen predictors, while the Ackers-White predictor is more sensitive with the velocity. The change is more significant for higher velocity (greater than 0.70 m/s), which corresponds to a Froude number of 0.24. In irrigation canals the Froude number normally varies from 0.2 to 0.5.

Effect of B-h ratio and side slope

For the evaluation of the effect of change in bed width and side slope on the sediment transport capacity the discharge, bed slope and the roughness was kept constant. For each side slope, the bed width was changed to get different B-h ratio. Change in bed width also changed the flow velocity. The results for three different side slopes of 1:1, 1;1.5 and 1:2 are presented in Figure 4.19. Engelund-Hansen and Ackers-White show similar trends of predictions. For a low B-h ratio the transport capacity is also low and the capacity increases with the increase in the B-h ratio. After a certain B-h ratio the transport capacity almost remains constant for the Engelund-Hansen method while it decreases in the Ackers-White method. In casa of Brownlie, the transport capacity is the highest for the lowest B-h ratio and almost constantly decreases with the increasing B-h ratio.

For all the three predictors, the flatter the side slope the lower is the transport capacity. The reason is that for a given B-h ratio, the steeper side slope increases the flow depth and the flow velocity. Hence, the transport capacity increases for steeper side slopes.

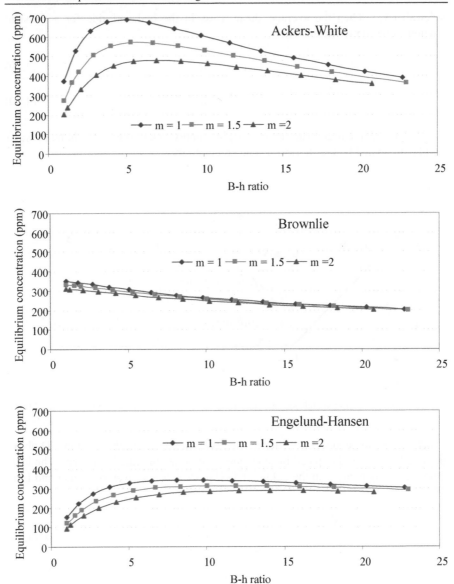

Figure 4.19   Effect of B-h ratio and side slope in the prediction of equilibrium concentration.

4.6.2   Velocity distribution in a trapezoidal canal

The velocity distribution in a trapezoidal canal is influenced not only by the effect of boundary shear, but also by the changing water depth along the slope. If the canal is not very wide, then this influence is significant and the concept of uniform velocity across the section cannot be used. Moreover the roughness is not constant along the perimeter due to the presence of bed forms and protection works or vegetation in the sides. Einstein (1942) suggested that the total area of such canals can be divided into that corresponding to the bed (A_b) and that corresponding to the sidewall (A_w). Then the average shear stress in the bed and side for a constant friction slope can be written as:

$$\tau_b = \rho g R_b S \tag{4.39}$$

$$\tau_w = \rho g R_w S \tag{4.40}$$

This lead to the argument that the surplus energy in any flowing volume of water will be dissipated by the shortest distanced boundary.

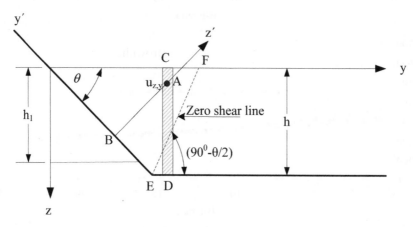

Figure 4.20   Definition sketch for calculation of depth averaged velocity (Yang, et al., 2004).

Based on this concept the surplus energy at any point in the column CD above the line EF will be dissipated by side slope and below the line by the bed. The local velocity at any point A in the column CD (Figure 4.20) is given by (Yang and Lim, 1997):

$$\frac{u_{(z',y')}}{u_*} = 2.5 \ln \left(f \frac{z'}{z_0} \right) \tag{4.41}$$

with
$$f = \frac{u_{*(y)}}{u_{*(h)}} \tag{4.42}$$

where

$u_{(z',y')}$ = velocity at points in the shaded column

u_* = overall mean shear velocity (m/s)

$u_{*(y)}$ = local shear velocity based on local boundary shear stress (m/s)

$u_{*(h)}$ = local shear velocity at the centre of the canal (m/s)

z_0 = k/30 for a rough boundary, where k is roughness height

To compute the depth averaged velocity over the column CD the following assumptions are made:
- the roughness in the bed and sides are equal, so the line of zero shear (EF) is the bisector of the angle between bed and slope;
- the local shear velocity can be replaced by the local average shear of either bed or side.

Now integrating equation (4.41) along the column CD the relation for the depth averaged velocity ($\bar{u}_{(y)}$) in a stream column is given by (Yang, *et al.*, 2004):

$$\frac{\bar{u}_{(y)}}{u_*} = 2.5\ln\left(f\frac{h_1}{z_0}\cos\theta \right) - 2.5(1+\beta) \qquad (4.43)$$

In non-wide canals and near the side wall in wide canals the maximum velocity is located below the free surface, which is known as dip phenomena. The second term in the RHS of equation (4.43) accounts for the dip phenomena and for a smooth canal the value of β is given by:

$$\beta = \frac{1.3}{\sin\theta}\exp\left(-\frac{y}{h_1} \right) \qquad (4.44)$$

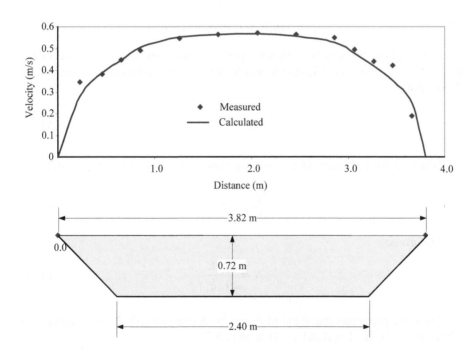

Figure 4.21 Calculated and measured velocity distribution in one of the sub-secondary canals (SS9F).

The velocity distribution in some of the sub-secondary canals (trapezoidal, earthen) of SMIS, Nepal was measured and for the rough canals the coefficient β has been found to be:

$$\beta = \frac{0.8}{\sin \theta} \exp\left(-\frac{y}{h_1}\right)$$
(4.45)

where

h_1 = water depth at point 1(m)

θ = angle made by side slope with the water surface

Figure 4.21 shows the predicted velocity distribution by equation (4.43) and the measured one in the sub-secondary canal S9F that has been found to be matching quite satisfactorily.

4.6.3   Exponent of velocity in sediment transport equation

Sediment transport predictors are of different forms and complexities depending upon the assumptions and the basic approaches used in the derivation. There is no general agreement on the type of variables that are required to define the sediment transport, but the most frequently used ones are:

$$q_s = f(u, h, S, \rho, v, \rho_s, d_{50}, g, \sigma_g)$$
(4.46)

For simplicity the sediment transport per unit width can be approximated by the power law, where the coefficients M and N are supposed to be constant locally (De Vries, 1987) as:

$$q_s = MV^N$$
(4.47)

Differentiation of sediment transport equation (4.47) with respect to V results:

$$\frac{dq_s}{dV} = MNV^{N-1}$$
(4.48)

This gives:

$$N = \frac{dq_s}{dV} \frac{V}{q_s}$$
(4.49)

For some predictors the value of N can be derived directly by comparing with equation (4.47), like Engelund and Hansen (1967):

$$q_s = \frac{0.05 \, V^5}{(s-1)^2 \, g^{0.5} \, d_{50} \, C^3}$$
(4.50)

which gives N = 5.

Equation (4.49) can be used for more complex predictors like Ackers and White (1973), Van Rijn (1984a,1984b) and Brownlie (1981) (Klaassen, 1995). The Ackers and White predictor for sediment transport is given by (ref subsection 3.4.2):

$$q_s = G_{gr} V\, d_{35}\, (\frac{V}{u_*})^n \tag{4.51}$$

Differentiating and comparing the results with equation (4.49) gives the following relation for the exponent N (De Vries, 1985):

$$N = 1 + \frac{m' F_{gr}}{(F_{gr} - A)} \tag{4.52}$$

where, the dimensionless mobility parameter (F_{gr}) is given by equation (3.70).

Similarly Brownlie's predictor for sediment discharge is given by (ref subsection 3.4.2):

$$q_s = \frac{0.007115\, q}{s}\, c_f\, (F_g - F_{gcr})^{1.978}\, S^{0.6601}\, (\frac{R}{d_{50}})^{-0.3301} \tag{4.53}$$

Differentiating and comparing with equation (4.49) give:

$$N = 1 + \frac{1.978 F_g}{F_g - F_{gcr}} \tag{4.54}$$

Assessment of N (Klaassen, 1995, Méndez, 1998) shows that it depends upon the flow conditions and sediment characteristics. In most cases (except Engelund and Hansen, Bagnold (1966)) it increases with a decrease in flow velocity and decreases in particle size (Figure 4.23). For a higher velocity the value of N remains fairly constant for a given particle size.

In case of Brownlie's predictor the exponent N is independent of bed slope, sediment size and geometric standard deviation for grain Froude numbers (F_g) more than 10, which refers to the velocity slightly more than that required for the initiation of motion (Figure 4.23). Moreover the value of N is never less than 3 for sediment size smaller than 0.50 mm and Froude number smaller than 0.6.

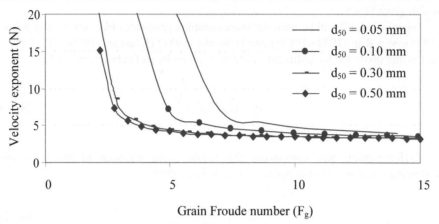

Figure 4.22   N versus grain Froude number for Brownlie's predictor.

The exponent N is more sensitive to sediment size in case of the Ackers-White predictor (Figure 4.23). Near the initiation of motion the value of N cannot be determined. N is always more than 4 for Froude number less than 0.6, which is also the upper limit of the normal flow conditions in irrigation canals.

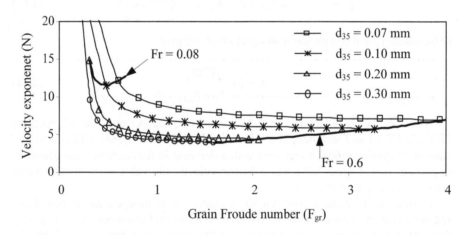

Figure 4.23   Variation of N with grain mobility parameter in Ackers-White predictor.

Moreover F_{gr} (in the Ackers and White method) is also related to the water depth and in case of a trapezoidal canal water depth also changes in side slopes in addition to the velocity. However calculations (Figure 4.24) show that the change in water depth has a negligible effect on F_{gr} and hence can be neglected.

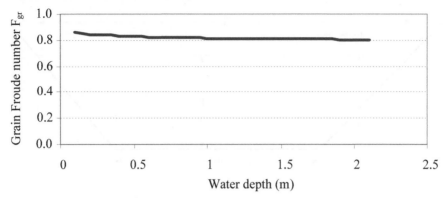

Figure 4.24   Variation of F_{gr} with water depth.

4.6.4   Correction factor

If the whole cross section of the canal is divided into small columns of finite width dy (Figure 4.25), then sediment transport through a stream column of width dy and local depth averaged velocity u is given by:

$$q_s = Mu^N \qquad (4.55)$$

Now the summation of all the stream tube transports will give the sediment transport rate of the whole section.

$$Q_s = \int_T Mu^N dy \qquad (4.56)$$

here T is the top width of the canal and dy is the width of the stream tube.

Using the variables averaged over the whole cross section, the sediment transport through the whole section becomes:

$$Q_s = B_s MV^N \qquad (4.57)$$

where B_s is the sediment transport width of the section.

The sediment transport given by equation (4.57) is not equal to that given by equation (4.56) and the difference is due-to the nonlinear relationship between the sediment transport and the velocity. Let α_s be the difference in the total transport rate computed by using the two methods then:

$$\alpha_s = \frac{\int_T Mu^N dy}{B_s MV^N} \qquad (4.58)$$

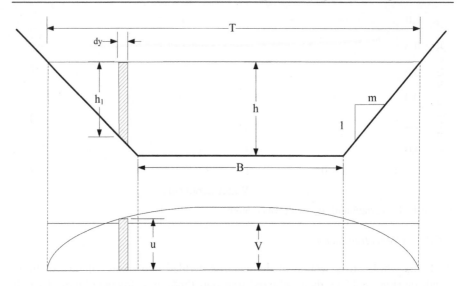

Figure 4.25   Schematized stream column and velocity distribution in a non-wide canal.

Here α_s is the correction that should be applied to account for the influence of the non-uniform velocity distribution in trapezoidal and non-wide canals. In case of rectangular canals there will be no influence in velocity distribution due to the side slope, still the non-uniformity in velocity will exist especially if the B-h ratio is small.

Use of simplified equations is useful to understand the effect of the shape on sediment transport behaviour, but for the actual adjustment in the predictor they should be analysed separately. The coefficients M and exponent N have different values and depend upon different variables in different predictors. Hence the correction in the prediction for changing side slope, B-h ratio, sediment size and the velocity exponent (N) for each predictor should be evaluated separately. In this study three total load predictors, namely Ackers-White, Brownlie and Engelund and Hansen have been evaluated.

For the evaluation, the following ranges of hydraulic and sediment characteristics were used:
– Froude number 0.05 to 0.5;
– sediment size (d_{50}) 0.075 to 0.5 mm;
– bed width to water depth ratio 2 to 12;
– side slope 1:0 to 1: 3;
– Chézy roughness coefficient 35 to 60;
– number of stream tubes 40;
– geometric standard deviation of 1.2 to 1.8.

Engelund and Hansen predictor

In case of the Engelund and Hansen predictor the exponent N is constant and is taken equal to 5. For this predictor also the correction is a function of *B-h ratio* and side slope. The corrections are given by:

– trapezoidal canal:

$$\alpha_s = 1.2785 \left(B/h \right)^{-0.0937} m^{0.078}$$ (4.59)

– rectangular canal:

$$\alpha_s = 1.2 \left(B/h \right)^{-0.0663}$$ (4.60)

Ackers-White total load predictor

In the Ackers-White total load predictor the correction factor α_s is a function of B-h ratio, velocity exponent (N), side slope (m) and sediment size (d_{50}). In the analysis the sediment size (d_{50}) is replaced by the dimensionless grain parameter (D_*) and the exponent N varies from 3 to 10. All the other hydraulic and sediment characteristics are same as that for the Engelund and Hansen predictor. Using non-linear regression the correction for the ranges of flow conditions and sediment characteristics as indicated above is given by:

– trapezoidal canal:

$$\alpha_s = 0.396 \left(B/h \right)^{-0.1012} N^{0.7514} m^{0.0541} \left(\log D_* \right)^{0.2427}$$ (4.61)

– rectangular canal:

$$\alpha_s = 0.0868 \left(B/h \right)^{-0.1699} N^{1.3175} D_*^{0.3153}$$ (4.62)

Brownlie's predictor

In case of the Brownlie's predictor the correction factor is a function of bed width to water depth ratio, velocity exponent (N) and side slope (m) of the canal. The exponent N varies from 3 to 6. All the other hydraulic and sediment characteristics are same as that for the Engelund and Hansen predictor.A non-linear regression of the computed results yields the following relationships for correction:

– trapezoidal canal:

$$\alpha_s = 1.023 \left(B/h \right)^{-0.0898} N^{0.1569} m^{0.078}$$ (4.63)

– rectangular canal:

$$\alpha_s = 0.8492 \left(B/h \right)^{-0.0361} N^{0.2106}$$ (4.64)

4.6.5   Predictability of predictors with correction

Three procedures are defined for the comparison of the sediment transport computations:

–   *procedure I*. In this procedure the hydraulic radius (R) is taken as representative variable for the water flow and the average width is taken as the representative width of the canal;
–   *procedure II*. In this procedure the water depth is taken as representative variable and the total sediment transport is calculated using bed width (B);
–   *procedure III*. In this procedure the sediment transport capacity of each stream tube is computed and added for the whole area, which can be written as:

$$Q_s = \int_T q_{s,y} \, dy \qquad (4.65)$$

The total sediment transport in the section is also given by:

$$Q_s = \alpha_s q_s B \qquad (4.66)$$

where q_s is the sediment transport rate per unit width calculated from the mean flow velocity and other variables as indicated in the respective predictors. So, the transport rate in this procedure is the quantity obtained from the predictors taking the canal as wide and then making a correction for non-wide conditions and side slope.

Table 4.4   Selected data set for the evaluation of sediment transport capacity computation procedures.

Investigator and year	Data code	No. of records
Gilbert, G.K. (1914)	GIL	12
U.S. Waterway Experiment Station (1935A)	WSA	35
U.S. Waterway Experiment Station (1936B)	WSS	17
Barton, J.R. and Lin, P.N. (1955)	BAL	9
Nomicos, G. (1957)	NOM	5
Vanoni, V.A. and Brooks, N.H. (1957)	VAB	4
Laursen, E.M. (1958)	LAU	6
Vanoni, V.A. and Hwang (1965)	VAH	6
Guy, H.P. et al (1966)	GUY	8
Government of Pakistan (1966-69)	EPB	20
East Pakistan Water and Power (1967)	EPA	21
Franco, J.J. (1968)	FRA	7
Pratt, C.J. (1970)	PRA	8
Davies, T.R. (1971)	DAV	13
Onishi, Jain and Kennedy (1972)	OJK	4
Nordin, C.F. (1976)	NOR	26
Sony, J.P. (1980)	SON	5

For the comparison of the three procedures, the selected data set for total load transport form Brownlie (1981) compilation of 55 flume and 24 field data sets was considered. The selection criteria are based on the flow conditions and sediment characteristics that normally exist in irrigation canals. The selection criteria are:
- sediment size less than 0.5 mm;
- Froude number less than 0.5;
- *B-h ratio* less than 8;
- sediment concentration 100 - 1500 ppm;
- geometric standard deviation of bed particle size (σ_g) less than 1.5;
- type of bed form dune or less;
- data that have all the measurements required for sediment transport computation.

A total of 149 data sets were selected out of the total 1,049 flume data sets of Brownlie's compilation (Table 4.4). All the data are from rectangular flume with a maximum width of 2.4 m. Hence the width for the sediment transport in the above mentioned procedures will be equal to the bed width. Figure 4.26 shows the sediment size and concentration range of the selected data.

Three predictors, namely Brownlie, Ackers and White and Engelund and Hansen are used for the prediction of the total load under equilibrium condition. The three procedures have been compared on the relative basis. The measured sediment transport is compared with the predicted sediment transport calculated by using one of the three procedures. Then the predictability of each procedure is measured on the basis of well-predicted values within certain accuracy range. For a given error factor (*f*) the upper and lower range is given by:

$$\frac{\text{measured value}}{f} \leq \text{predicted value} \leq \text{measured value} * f \qquad (4.67)$$

$$\text{accuracy} = \frac{\text{number of well-predicted values}}{\text{total values}} \qquad (4.68)$$

The comparison results (Figure 4.20) show that the predictability of the predictors is improved when a correction is applied to incorporate the effect of B-h ratio (procedure III). The predictability is improved when hydraulic mean depth (R) is used in place of water depth (h) for Brownlie's predictor. For Ackers-White, the use of water depth (h) produced better accuracy as compared to the use of hydraulic depths (R). For Engelund-Hansen, there is no significant change in the prediction accuracy for both cases (procedure I and procedure II). The use of representative width as bed width or average width (average of bed width and water surface width) has no influence in the result as all the data were from rectangular section. However, in trapezoidal canal sections the method of taking representative width should have significant effect in the sediment transport volume. The most logical option could be to use the average of the bed width and the top with of the canal and in this research the average width has been used as the representative width.

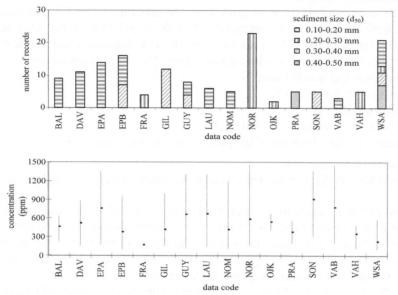

Figure 4.26   Characteristics of selected data.

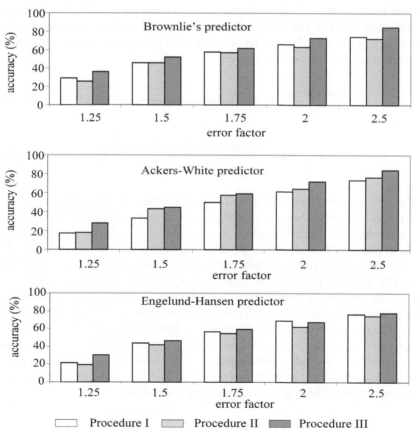

Figure 4.27   Sediment transport prediction accuracies of the three procedures.

4.7   Conclusions

The sediment movement is the function of hydraulic and sediment characteristics. The proper estimation of design discharge and the roughness helps to predict the hydraulic conditions more accurately. An understanding of the influence of canal geometry in the sediment transport capacity of the canal for a given discharge will assist in the proper selection of bed width, bed slope and side slope of the canal during the design.

In general the reliability of sediment transport predictors is low and at best they can provide only estimates. A probable error in the range of 50-100% can be expected even under the most favourable circumstances (Vanoni, 1975). The error is expected to increase further if the calculations are based upon average values of flow and sediment parameters. Engelund and Hansen (1967) have pointed out that even after the theoretical side wall correction, the experimental flume data should be considered with caution. It is because the shear is dependent not only on the relative roughness of bed and walls but also on the width to depth ratio of the flume. Moreover Brownlie (1981) mentions that the field data have slightly higher sediment concentrations than laboratory data for a similar ranges of dimensional groups.

Even though there is improvement in the accuracy of prediction while the corrections are applied to the sediment transport predictors as compared to the conditions of without prediction, accuracy of prediction is still low. The accuracy is less than 50% for a error factor of 1.5. One predictor predicts well for one data set, while for the other data set the predictability if very poor. This clearly indicates the limitations of the available predictors for sediment transport. They have been derived from the limited range of hydraulic and sediment characteristics and their use outside those ranges should be made with care. Predictors should be tested with the field conditions where they are to be used and adjustment should be made if found necessary.

5
Towards a new design approach

5.1   General

For countries like Nepal, with relatively young and fragile mountains, high intensity rainfall patterns, agricultural activities on the sloping and marginal lands and rapidly shrinking forest areas, the sediment is an integral part of the river flows, which is difficult to be excluded by the sediment removal arrangements in the headworks. Hence, irrigation canals have to carry the sediment loads. The concentration of sediment and its variation during the irrigation season depends upon the type and efficiency of sediment removal facilities.

While preparing the design of canals from sediment transport aspect for a new irrigation scheme or for the modernization of an existing scheme, the information on the inflow concentration and type of sediment and its variation in time should be carefully worked out. In many irrigation schemes, the sediment removal options provided have become useless, partly due to design shortcomings but mostly due to operational requirements that are difficult to implement. So, a theoretically sound and effective sediment removal/control technique may not be equally effective for all the schemes, due to their specific characteristics and operational limitations. This aspect should also be analysed while designing the canal, that is; how the system will behave if the assumptions on sediment removal at the headwork are not met.

Once the sediment enters into the scheme then the management efforts should be to transport or deposit the sediment in a controlled manner such that the functional objectives of the schemes are not hampered. Hence the design of the canal for sediment transport should be an integrated approach of hydraulic calculations that take care of well defined management plans. In this chapter a brief description of the canal design methods that are in use in Nepal and more specifically in Sunsari Morang Irrigation Scheme (SMIS) is given, then the proposed design approach in terms of hydraulic calculations and the management aspect that should be taken into consideration is presented and at the end the discussion on the proposed design method is given.

5.2   Hydraulic aspects of canal design

The purpose of the hydraulic design of canals is to determine the flow depth required to pass a known discharge for a given geometry and roughness. Geometry of a canal is defined by the bed width, side slope and bed slope. Hence, theoretically any of the three parameters, i.e., bed width, side slope or bed slope can be varied to obtain a desired depth for a given discharge. Since, it is not practical to construct a canal with varying side slope along its length, it is normally taken constant unless the canal discharge is reduced significantly and steeper slopes are possible to adopt. Hence, for a given condition (soil type, discharge) the side slope is selected and taken as constant, while bed width and bed slope are taken as the design parameters.

Similarly, roughness depends upon the flow conditions as well as on other factors like type of construction material, quality of construction and/or maintenance, canal shape, geometry and vegetation. Although the change in roughness will change the water depth for a given canal geometry and discharge, it is not taken as design parameter. Roughness is related qualitatively to the type and quality of the construction as well as the maintenance work, but quantitatively it is not practical to specify a roughness and ask to construct or maintain a canal accordingly.

5.2.1   Existing canal design approaches in Nepal

Based on design considerations, irrigation canals in Nepal can be broadly divided into two groups, canals in the hills and canals in the Terai (alluvial plain in the southern part). Hill canals:
- *are relatively smaller in size*. In most cases the command area is less than 1,000 ha;
- *have limited space for the width*. The canals are normally designed with smaller B-h ratio and steeper side slope. In some cases rectangular lined canals are cheaper than trapezoidal earthen canals;
- *pass through relatively hard soil*. It is possible to provide steeper bed slopes;
- *relatively coarse sediment*. The sediment load coming into the hill canals is normally coarser. Sediment transport problems in hill canals are being tackled efficiently by using a sediment trap and providing steeper longitudinal slopes.

In general the hill canals are narrower in width and steeper in slope, while the Terai canals should have a gentle bed slope to avoid the loss of command area. Moreover, Terai canals receive more fine sediment that is difficult to separate and pass through erodible soil. Hence more emphasis has to be given in the design of Terai canals for tackling sediment transport issues.

The problem of sediment in irrigation schemes in Nepal is recognized and accordingly for the design of unlined earthen canal in the hills and plains different methods have been suggested. In the design manuals for the design of irrigation canals in Nepal, generally two conditions have been identified (Department of Irrigation, 1990a):
- *water flowing without sediment in unlined canals*. For sizing the canals Manning's equation is recommended while for limiting the slope to prevent bed erosion tractive force equations are recommended;
- *water flowing with sediment in unlined canals*. In this case the manual suggests to satisfy both the non-scouring and non-silting criteria. For sizing the canals it is suggested to use Manning's equation for all canals in the hills and small canals in the Terai, while Lacey's regime formula or White-Bettess-Paris tables (White, *et al.*, 1981b) for large canals in the Terai. For limiting the slope or preventing erosion in the bed tractive force equations are suggested. For computing the sediment transport capacity Engelund and Hansen or Ackers and White equations are suggested.

The manuals recommend that the use of Lacey's equations should be restricted to the areas where the sediment size and concentration is expected to be similar to those implicit in the formulae. It is recommended that for large canals, a more thorough assessment of the sediment balance is carried out using quantitative formulae. No specific standards have been setup so far, hence the constants and coefficients in the empirical equations are selected on personal judgement.

5.2.2   Canal design methods in SMIS

Similar to the many large scale irrigation scheme canals, the canal system of SMIS was also designed using Lacey's regime equations. This irrigation scheme faced severe sediment deposition problems and during modernization the sediment transport aspect was also given due importance in the design of canals. However, in the absence of clear and defined guidelines for the design, different approaches have been used. During earlier phases (stage I and II) of modernization, Lacey's regime method (Method I) was used while later on (stage III) the tractive force method with energy concept for preventing deposition was used (Method II) (Department of Irrigation, 1987, 2003). A brief description of these methods and their suitability will be discussed in the following sub-headings.

Method I

The set of Lacey's regime equations used in the design and the values of the constants taken are (Department of Irrigation, 1987):

$$V = \left(\frac{Qf^2}{140e^2} \right)^{\frac{1}{6}} = 0.483Q^{\frac{1}{6}} \tag{5.1}$$

$$D = 0.525 \left(\frac{Q}{e^2 f} \right)^{\frac{1}{3}} = 0.636Q^{\frac{1}{3}} \tag{5.2}$$

$$B = 0.8B_s = 2.9\sqrt{Q} \tag{5.3}$$

$$B_s = 4.83e\sqrt{Q} \tag{5.4}$$

$$S_f = \frac{0.00303e^{\frac{1}{3}} f^{\frac{5}{3}} E}{Q^{\frac{1}{6}}} \tag{5.5}$$

$$E = \frac{P}{B_s} = 0.8 + 0.2 \left(\frac{\sqrt{1+x^2}}{x} \right) \tag{5.6}$$

and x is given by

$$x = 0.1 B_s / D = 0.57Q^{1/6} \tag{5.7}$$

where

B_s = water surface width (m)

B   = bed width (m)

D   = design water depth (m)

E   = shape factor

e   = width factor (taken 0.75 in SMIS)

f   = silt factor (taken 1.0 in SMIS)

P   = wetted perimeter

Q   = design discharge (m^3/s)

S_f = water surface slope (m/m)

V   = mean flow velocity (m/s)

The regime theory postulates that for a given discharge, sediment diameter and concentration the width, depth, mean velocity and slope of a sediment transporting canal are uniquely determined. The statement is subject to the provision that the sediment is loose, incoherent, the channel is active and the bed is in movement. That is, there is no restriction to the formation of regime conditions. The major issues in the use of Lacey's regime equations are:

– *implicit use of sediment parameters.* The sediment size and concentration appear implicitly in the equations. Only the silt factor appears in the equation that is related to mean sediment size (d_{50}). There is a common consensus among the researchers that more number of hydraulic and sediment related parameters, other than the sediment size and discharge, have influence in the sediment transport process. The disagreement, however, still exists on how much influence one parameter will have over the others. That is the reason, the sediment transport capacity predicted by the available predictors varies by a large margin for the same hydraulic and sediment related parameters. In the regime equations, all sediment related variables are included in a single parameter, the silt factor;

– *concept of incoherent alluvium and bed width.* Most of the irrigation scheme in Nepal pass through the terrain that can not be considered to have unlimited envelope of sediment identical to the transported sediment as assumed in the regime theory. The regime width is constant for a given discharge and is independent of the type of bank material. In practice, even the natural rivers with strong banks are narrower than the ones with erodible banks. Hence, to achieve the regime conditions, the banks should not offer any resistance. Such conditions are difficult to find in the irrigation canals in Nepal;

– *introduction of flow control structures.* For a canal to adjust its slope and width and attain a regime condition an incoherent perimeter and a long reach without any restriction to flow is needed. Such conditions are difficult to find in modern irrigation canals. Even the old irrigation schemes, designed on a supply based concept, are being converted to a demand based concept during modernization. As a result, more of flow control structures are being added to regulate the flow to meet the changing demand. Hence, there is hardly any possibility of attaining the so called regime condition in present day irrigation canals;

– *changing water flow.* Regime conditions assume canals to carry relatively constant flows with little variation. The modern irrigation canals are highly demand based and the discharge varies constantly within the irrigation season. It will be difficult to assess sediment transport process in such canals using the regime theory;

- *roughness implicit in other equations.* The theory assumes that in a self formed channel in loose sediment the roughness is implicit in the values of the hydraulic mean depth and the slope it adopts. It is not clear how to compute the roughness of a canal that is not in regime. Normally the irrigation canals have defined boundary and the side slope and different protection and repair works are done to maintain the side slopes that will definitely affect in the equivalent roughness of the canal section. The effect is more pronounced in case of non-wide canals.
- *more equations.* As discussed above the hydraulic design of a canal is basically to find the three variables; bed width (B), water depth (h) and bed slope (S_0). Hence, three equations are needed to find a unique solution. But, Lacey's regime theory provides more than three equations. This creates confusion which of the three equations should be taken in the design.

No studies have been made on how much deviation there is between the predicted and actual stable canal parameters in Nepal. However, in India the predicted values deviated from actual stable canal values by 11 to 84%. Some adjustments have been made in the regime equations to make them compatible with the local conditions. Examples of such equations are that of Chitale's best fit equations (Chitale, 1966) and the Irrigation Research Institute, Roorkee equations for northern India (Varshney, *et al.*, 1992).

The design manuals of the Department of Irrigation have also recommended to use the equations only in the conditions that are similar to the conditions for which the equations were derived. But there are so many variables involved in the sediment transport process and the Lacey's regime equations use so little parameters that it becomes difficult to make any tangible comparison, between the conditions of the place of interest and the conditions where these equations have been claimed to be successful.

Despite all the limitations in the equations the design engineers use the Lacey's regime equations for the design of canal mainly because they are simple and easy. Design discharge and the mean sediment size are the only information needed to start the design.

Method II

The second approach used is based on the energy concept that states that the sediment transport capacity (stream power) should be constant or non-decreasing in the downstream direction. The basic principle of this method is that any sediment entering into the system should be transported to the end without deposition in between. For the control of erosion in the bed the shear stress is restricted to a certain safe limit.

As mentioned above three equations are needed to solve the three independent variables of canal design i.e., bed width (B), bed slope (S_0) and water depth (h). In this approach one of the three resistance equations (Chézy, Manning or Strickler) makes the first equation.

The bed width to water depth ratio (B-h ratio) makes the second equation. Different empirical relationships exist for determining the B-h ratio and side slope. Side slope is mostly based on the soil type while the B-h ratio depends upon various factors as economy, sediment transport, seepage and maintenance. In SMIS the values as shown in Table 5.1 have been used.

Table 5.1   B-h ratio and side slope for different discharges.

Flow range (Q m³/s)	V:H	B-h ratio
0.1 to 1.0	1:1	1 to 4
1.0 to 15.0	1:1.5	2.7 to 7
15.0 to 100.0	1:2	5.5 to 10

Check for erosion. Now the bed slope is assumed and the two equations are used to find the water depth and bed width. The slope is then checked for erosion using the following condition:

$$\tau_b = \rho g h S_0 \qquad (5.8)$$

where

g = acceleration due to gravity (m/s²)
h = water depth (m)
S_0 = bed slope (m/m)
ρ = density of water (kg/m³)
τ_b = maximum tractive force (<3-5 N/m²)

Check for suspended load. For controlling the deposition of suspended sediment, the Vlugter (1962) energy concept has been used. It states that sediment particles will be transported in any concentration by the flowing water when the fall velocity (w) is less than a certain threshold, given by:

$$w_s \leq \left(\frac{\rho_w}{\rho_s - \rho_w} \right) V S_0 \qquad (5.9)$$

Shoemaker (1983) used this concept and proposed the concept of stream power for designing a canal with suspended sediment. The stream power is given by:

$$E = \rho g V S_0 \qquad (5.10)$$

where

E = stream power (W/m³)
S_0 = bed slope (m/m)
V = mean velocity (m/s)
ρ = density of water (kg/m³)

The energy of the main canal at the head of the off-take is computed and the off-take canal is designed such that it would have energy at least equal to or more than the energy of the main canal to ensure no sediment deposition. The criteria used to test this condition are:

$$V S_0 = \text{constant or non-decreasing} \qquad (5.11)$$

Check for bed-load. The relative transport capacity of flowing water can be written as:

$$\frac{T}{Q} \propto \frac{Bh^3 S_0^{\ 3}}{Bhh^x S_0^{\ z}} \qquad\qquad (5.12)$$

where

B = bed width (m)

h = water depth (m)

S_0 = bed slope (m/m)

T = sediment transport rate (m^3/s)

T/Q = relative transport capacity

Q = discharge (m^3/s)

x, z = exponents depending upon the choice of sediment transport and water flow equations

Using Vlugter (1962) concept and Strickler's equation for water flow the necessary and sufficient condition for non silting due to bed-load is (Dahmen, 1994):

$$h^{\frac{1}{2}} S_0 = \text{constant or non-decreasing in downstream direction} \qquad (5.13)$$

This criterion is used to check the possibility of sediment deposition in the bed. If the criterion is not satisfied then the slope is changed. If, however, the slope cannot be changed due to the area to be commanded, the existing off-take or other canal structures then the new calculation is started using a different B-h ratio.

Comparatively, the energy approach is better as compared to the regime equations as it uses a more rational approach for including sediment transport aspects in the design. Firstly the sediment transport capacity (T m^3/s) is determined based on the type and concentration of the sediment expected to be transported. Then for a given discharge (Q m^3/s) relative transport capacity T/Q is computed which is assumed to be proportional to the energy of the flowing water (equation 6.8). The design philosophy is that, if the energy of the off-taking canal is kept equal to that of the parent canal, then the suspended sediment load that the canal receives from the parent canal can be transported in suspension. And if the transport capacity of the off-taking canal is maintained equal or non-decreasing in the downstream direction then there will be no deposition in the canal. In this way all the sediment is transported to the desired location.

The advantage of this approach is that all the canals downstream of a point under consideration are related with each other. The secondary canal's transport capacity is determined by the transport capacity of the main canal near the in-take of the secondary canal. Similarly, the capacities of the sub-secondary canals are related to the capacity of secondary canal and so on. If due to some reasons, a canal can not be designed with the energy comparable to that of the parent canal then the designer will have prior knowledge that the canal might have some sediment transport problems in that canal and can suggest some operational or management solutions.

There are, however, some limitations of this approach in the design of canals that carry appreciable amount of sediment:

− *sediment transport rate only a function of velocity (V) and bed slope (S₀).* As discussed in section 4.4, irrigation canals differ from rivers due to the limited B-h ratio and the presence of side slopes. The majority of irrigation canals are non-wide and trapezoidal in shape with the exception of small and lined canals that may be rectangular. In a trapezoidal section the water depth changes from point to point in the section and hence the shear stress. The effect would be more pronounced if the bed width to water depth ratio (B-h ratio) is small. The change in velocity distribution in a canal in view of the change in boundary shear and water depth along the cross section will influence the sediment transport capacity also. Hence, by relating the transport capacity with the velocity and bed slope only, the major characteristics of an irrigation canal are ignored and the sediment transport process is not completely described;

− *non-decreasing energy not the sufficient condition.* The sediment transport predictors are more sensitive to velocity as compared to the bed slope. Taking other variables constant the exponent of velocity is equal to 5 for Engelund-Hansen, greater than 3 for Brownlie and greater than 4 for Ackers-White predictors. The slope parameter does not appear explicitly in Ackers-White and Engelund-Hansen predictors while in Brownlie the exponent to slope is 0.6. Hence, the transport capacity may not change proportionally with change in VS_0 for sediment load outside the De Vos assumptions (fine sediment of size less than 70 μm). Discharge in irrigation canals decreases in downstream direction, hence the roughness increases for the same canal shape and bed material size and the velocity (V) decreases. The product VS_0 can be maintained constant or non-decreasing by three different methods:

o by changing the bed width while keeping the bed slope constant;
o by changing the bed slope while keeping the bed width constant;
o by changing both the bed width and bed slope.

The practical difficulty lies on the correct selection of method/option to get the constant VS_0. Sometimes, the topography restricts the change in slope while sometimes the bed width has to be kept constant. Moreover, for a canal with the same constant energy achieved by three different methods, the actual sediment transport capacity of the canal will not be same. Hence, the non-decreasing energy criteria may be necessary but not sufficient for preventing deposition in the canals;

− *extrapolation of the method for larger sediment size.* The De Vos statement was for very *fine* sediment (d ≤ 50 − 70 μm) and the assumption that it will be equally applicable to larger sized sediment is not yet justified.

5.2.3   *Proposed canal design approach*

For the design of canals with sediment load a design approach has been suggested that takes into account the bed forms and the effect of bed width to water depth ratio on the roughness (Depeweg and Paudel, 2003). In the following part the process of design will be discussed. In the end the summary of design calculation and flow diagram for the same will be presented.

Design discharge and sediment concentration

The capacity of the canal should be such that the embankments are not overtopped or the canal network is not at risk, when maximum or peak water demand (Q_{peak}) is being met. Hence Q_{peak} should be taken as the design discharge.

The problem lies in the selection of the design sediment concentration, i.e., what is the sediment concentration that the canal should convey in equilibrium condition while carrying the design discharge. It is clear that the water and sediment inflow rate in any canal system is not constant. The peak water flow (Q_{peak}) may not correspond to the peak sediment concentration. If there is provision of a sediment removal facility then there will be a known upper limit of sediment concentration. The design concentration which will be called "*dominant concentration*" is that value, for which there will be least erosion and deposition after one crop calendar year.

The dominant concentration is not the expected peak concentration that the canal has to convey. It is a value that will result in no net erosion or deposition after one irrigation season. Hence depending upon the inflow discharge and sediment concentration, erosion during one part and deposition during another part of the irrigation season is allowed, thus balancing the net effect in one irrigation cycle. Moreover in this case the sediment concentration will be used explicitly to arrive at a stable design.

Roughness

Equivalent roughness of the canal section is computed considering the effect of the B-h ratio and side slope. The roughness values in the side slopes and in the bed are estimated separately and the equivalent roughness is computed (ref section 4.5). The computation procedure can be summarized as:
– *determination of preliminary roughness*. The preliminary roughness is estimated based on construction material and maintenance condition (ref section 4.5.3). This roughness is used for the first approximation of the hydraulic parameters;
– *prediction of bed forms*. The preliminary hydraulic parameters together with the representative sediment size are used to find the possible type and size of bed forms (ref section 4.5.2);
– *roughness in the side*. The equivalent roughness height in the side slopes (K_{sl}) is derived from the preliminary roughness value;
– *roughness in the bed*. The equivalent roughness height in the bed (K_{sb}) is the sum of roughness due to the grain and roughness due to the bed forms;
– *equivalent roughness of the section*. Equivalent roughness of the section is then computed using equation (4.30).

Side slope

The side slopes depend mainly on stability criteria. In Nepal, selection of side slopes is made on the basis of soil type as well as the canal discharge (Department of Irrigation, 1990a). Similarly Dahmen (1994) suggested the side slope on the basis of discharge (Table 5.2). A steeper side slope is possible in smaller canals as compared to larger ones for the same soil type. Hence in the selection of side slopes, mechanical property of soil, depth of canal, method of construction, level of ground water table and local practices need to be considered.

Table 5.2   Suggested side slope of a canal in Nepal and by Dahmen (1994).

Nepal		Dahmen	
Discharge	side slope	Discharge	side slope
m³/s	H : 1 V	m³/s	H : 1V
< 1.0	1.0	< 0.2	1.0
1 to 15	1.5	0.2 to 0.5	1.0 to 1.5
> 15	2.0	0.5 to 10	1.5 to 2.0
		> 10	> 2.0

B-h ratio

Various empirical relationships have been developed and are in use for the B-h ratio. The value for different existing irrigation canals in different parts of the world is presented in Figure 5.1. Dahmen (1994) suggest the following relationships for canals carrying a relatively low sediment load.

for $Q > 0.2$ m³/s $\qquad\qquad \dfrac{b}{h} = 1.76 Q^{0.35}$ $\qquad\qquad$ (5.14)

for $Q < 0.2$ m³/s $\qquad\qquad \dfrac{b}{h} = 1.0$ $\qquad\qquad$ (5.15)

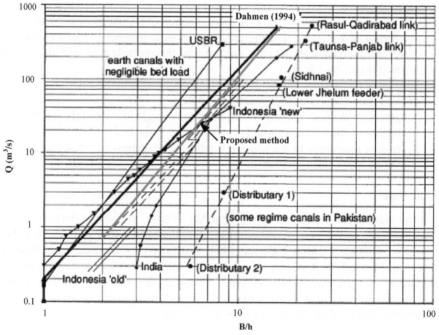

Figure 5.1   B-h ratio of existing canals of the world (after Dahmen, 1994) and proposed value.

Normally, the canals that carry sediment are wider than the ones that carry relatively clear water (Figure 5.1). In Nepal, the general practice is to use Lacey's regime width, but as discussed above the computed width results in a relatively wider canal since the side slope used is flatter due to soil stability considerations. Hence, for canals carrying sediment load and having a side slope comparable to that suggested in the design manuals in Nepal, the following relation for the selection of the B-h ratio is proposed based on the Indonesian practices.

For $Q < 0.5$ m^3/s $\qquad\qquad$ B-h ratio = 1 $\qquad\qquad\qquad$ (5.16)

For $Q > 0.5$ m^3/s $\qquad\qquad$ B-h ratio $= 2.5Q^{0.3}$ $\qquad\qquad$ (5.17)

Equilibrium sediment transport predictors

Total load predictors are proposed for predicting the sediment transport capacity under equilibrium conditions. The predictors by Brownlie, Ackers-White and Engelund-Hansen are some of them that can be used depending upon the expected sediment load and size. In case of Nepal, the sediment transport predictors by Brownlie and by Engelund and Hansen are proposed. The sediment transport capacity predicted is then corrected for the influence of the side walls of irrigation canals (ref section 4.6).

Calculation steps

The sequence of calculation for the design of a canal following the approach mentioned above is:
- selection of design discharge (Q), design sediment concentration (C) and sediment size (d_{50});
- determination of preliminary bed slope (S_0) and preliminary roughness (K_s). Preliminary bed slope is taken from the layout map of the canal;
- determination of B-h ratio and side slope (equation (5.14), (5.15) or Figure 5.1 and Table 5.2);
- using a preliminary bed slope and the B-h ratio, the water depth (h) is computed using any of the resistance equations (Manning, Chézy or Strickler);
- computed hydraulic parameters and sediment properties are then used to determine the type and size of bed forms and the equivalent roughness height in the bed (K_{sb}) (Van Rijn's method, equations 4.4 to 4.14);
- equivalent roughness of the canal section is computed taking initial roughness of the section as roughness in the sides (K_{sl}) and the roughness in the bed (K_{sb}) (equation 4.30 (Méndez, 1998));
- then the sediment transport capacity of the section under equilibrium condition is computed using one of the total load predictors and the value is adjusted for non-wideness of the canal section and the effect of the side slopes (equation 4.59 to 4.64);
- the computed equilibrium sediment transport capacity is compared with the expected sediment load that is to be transported by the canal section. If the carrying capacity of the section is not matching with the expected sediment load to be transported then the slope is adjusted and the computation is repeated.

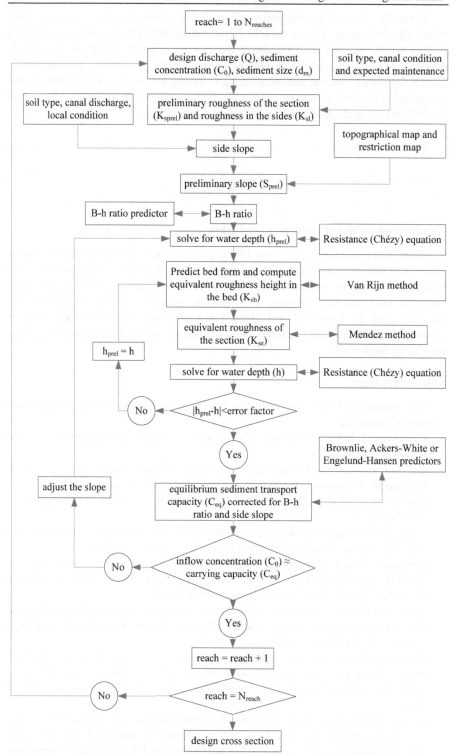

Figure 5.2   Canal design flow diagram.

A computer program DOCSET (Design Of Canal for Sediment Transport) has been prepared on the basis of this approach, which can be used to design a new canal as well as to evaluate the existing canals for their sediment transport capacities under equilibrium condition. The summary of the design process has been given in the flow diagram (Figure 5.2). The canal section designed in this way will be able to carry the design sediment load under normal flow conditions.

5.3   Management aspect of canal design

5.3.1   Operation aspect

In modern irrigation schemes the crop based water delivery plans are prepared in an attempt to increase the irrigation efficiency and crop production. Accordingly the canal network and the flow control systems are designed to make the system more flexible to meet the changing demand. From sediment transport aspect, flexibility means more control of the flow and hence more deviation from uniform flow conditions. That means the difference of actual sediment transport and the one predicted by the sediment transport equations becomes more.

Irrigation schemes that carry a sediment load need extra efforts to operate them properly. Two objectives have to be met simultaneously, firstly to supply the water as per the demand or as per the previously agreed schedule and secondly to ensure that the effects due to sedimentation/erosion are minimal. Delivery of water as per the demand requires adjustment in the water level and flow rate. The operation of control gates to manipulate the water level and discharge makes the flow in the canal system unsteady and non-uniform and will affect in the sediment transport behaviour. It is not possible to operate the scheme in a flexible way and also to reduce the sedimentation problem at the same time. Hence a compromise has to be made between the flexibility in water delivery and sediment deposition.

Management authorities have to follow strict operation rules to reduce the entry and deposition of sediment. If the rules are followed then the basic functional objectives of the scheme to supply water in sufficient quantity, right time and with fair share have to be compromised. If however, the rules are not followed properly, then the chances are that there will be unwanted deposition in some parts of the canal section, thus reducing the capacity. This will again lead to an inadequate and unfair distribution of available water. Hence the limitations of such irrigation schemes in terms of service delivery have to be understood at the time of design and accordingly operational plans should be prepared and used in the design.

5.3.2   Conveyance aspect

The objective of design of a canal for sediment transport should be to convey the incoming sediment load all the way down to the location where it can be disposed or deposited safely. But most design procedures assume each reach as an individual element. In such cases there will be no relationship between higher order and subsequent lower order canals in terms of their sediment transport capacity. If an off-taking canal has a smaller transport capacity than the parent canal then sediment diverted from the parent canal will be deposited near the head of the off-taking canal. If, however, the off-taking canal is designed with a higher transport capacity than the parent canal, the canal will always have more energy in the flow than necessary to transport the available sediment in the flow. Then there is possibility of bed erosion in the off-taking canal. The scheme can be operated optimally if the sediment transport capacities of the off-taking and parent canals are comparable.

The sediment management aspect should be fully understood and incorporated in the design. Simply designing a canal or reach of the canal for a certain transport capacity may solve the problem of that reach but from management aspect it is simply the shifting of the trouble from one place to another. It should be clearly indicated what is the objective in terms of sediment transport. If the objective is to transport the sediment down to the field then all the canals in the network should be capable of transporting the sediment load.

It has been observed that different levels of canals in the same system are designed using different approaches. For example in SMIS, secondary canals have been designed using Lacey's regime equations while sub-secondary and tertiary canals are designed using Manning's equation (Department of Irrigation, 1987). The design manuals also suggest to design the main and secondary canals using tractive force or regime concepts while lower order canals by simple Manning's or Chézy's equations (Department of Irrigation, 1990a). This gives an impression that the objective in the design from the sediment transport perspective is to transport the sediment up to the end of the secondary canals. Since the design approach does not include any specific plans and methods how the transported sediment is to be managed after that point, the operation of such schemes is difficult.

5.3.3   Provision of settling pockets

It is seldom possible to have the same slope for all the canals in an irrigation scheme. When there is restriction on slope then the B-h ratio is adjusted to arrive at the required sediment transport capacity. The adjustment in the B-h ratio alone may not be enough to design different canals with equal transport capacity. Similar conditions may arise even within a canal. Hence, one or more canals or one or more reaches of a canal may have more sediment deposition problems as compared to other for the given sediment load. The unequal sediment deposition pattern in different canals means that some farmers have to invest more as compared to others to receive the same amount of water. Besides, the uneven deposition of sediment along the same canal will put extra burden to the farmers downstream of that point to convey their share of water to the field. Normally the head reach farmer who is benefited due to the raised water level in the canal will not bother and participate in the removal of sediment in the canal. Thus the farmers of the tail end should have to invest more money for the canal maintenance and receive less water.

The settling basin at the head may become very large and expensive if it is designed to reduce the concentration to avoid deposition problems in all the canal network. Under such conditions, the provision of settling pockets may be helpful to reduce the social, operational and water distribution problem. In Nepal, generally the main and sub-secondary canals follow the contours, hence have flatter slopes than secondary canals that follow the ridges. Hence a settling basin at the head of a secondary canal will allow designing the canal with a flatter slope, thus even the high land within the command area can be provided with irrigation facilities.

5.3.4   Maintenance activities

Maintenance of a canal is primarily meant to restore the canal to the designed or planned state. In the process the level, cross sectional dimensions as well as the roughness of the section are corrected. Removal of deposited sediment from a canal network is the major maintenance activity of irrigation schemes carrying sediment laden water. Sediment deposition in the canal will have two fold influences. Firstly, the water diversion towards the canal is reduced due to the reduction in the driving head (difference in water level or in piezometric head between inflow and outflow cross-sections) and secondly the capacity is reduced due to the reduction in canal capacity.

If complete removal of the sediment is possible then there should be no problem, however, given the extent of deposited volume and available resource complete removal of deposited material is seldom possible. In case, when the sediment has to be removed partially then the desilting activity can be optimized if the causes of sediment deposition are understood and its effects in the operation and management of the irrigation schemes are explored. Improper planning and removal of sediment may not only increase the deposition rate but also not improve the water delivery. The Following criteria should be considered while making the maintenance plans:

− removal of sediment near the head reach increases the head between the source (parent canal or reservoir) and the canal and may increase the water drawing capacity. However, if the capacity of the downstream reach is not sufficient to transport the increased flow of water then backwater is created. This will again raise the water level near the head and the gain (in terms of driving head) made by desilting is balanced. Moreover the deeper canal section for the same flow rate will reduce the flow velocity and sediment deposition will take place at faster rate;

− sediment deposition starts from the upstream and continues towards the downstream reach of a canal if the incoming sediment load is more than the sediment transport capacity of the canal. If, however, the backwater effect is created due to the presence of a control structure or an obstruction and due to this the sediment transport capacity of the section is reduced than the sediment load in the flow, then the deposition will start from downstream end and move to an upstream direction. If sediment deposition near the intake is due to the backwater effect then removal of sediment may not help in increasing the discharge into the canal;

− if deposition is due to the lower transport capacity of the canal, then removal of sediment near the head reach may be beneficial to increase the water withdrawal into the canal;

− the irrigation policy of Nepal aims to operate the large scale irrigation schemes in joint management mode. Normally, the headwork, main canal and the secondary canals are under the responsibility of the government while the sub-secondary and the lower order canals are operated and maintained by the WUA. If lower order canals, which are the responsibility of the farmers, are not desilted then the desilting of the main and secondary canals is useless. The increased discharge in main and secondary canals can not be diverted to the lower order canals and only the head reach farmers will be benefited. Moreover, for a given discharge the water level becomes higher in the silted lower order canals, hence the water level in the parent canal has to be maintained higher. This will again create a backwater effect and a lower flow velocity. The deposition rate will be faster and the cleaned canals will be filled faster without any real benefits.

Disposal of sediment is also a major aspect that should be planned properly before the design. If the sediment does not have a fertilizing value then the design should aim to transport the sediment to the desired location where it can be deposited (controlled deposition) and removed without affecting the canal operation. Designing the canal network to transport all the sediment to the field is simply to transfer the problem from the canal to the field.

Figure 5.3   Desilting of canal is one of the major maintenance activities.

5.4   Modelling aspect of canal design

The design of a canal is a complex process that has to satisfy different operational requirements. Hydraulic calculations are based on a specific design discharge and sediment characteristics for uniform flow. The design information are derived considering the water requirement, sediment load in the river and the provisions of sediment removal facilities, expected maintenance conditions and the proposed operation plan. However, it is not possible that a canal designed for a specific water flow and sediment load to be non-silting and non-scouring for all the discharges and sediment concentrations. Since the water flow and sediment concentration keep on changing, it is unavoidable to have some deposition in one part of the irrigation season and some erosion in the other. Hence, the design should be able to produce a system that has a minimum net erosion/deposition at the end of the season. Hence, the design values of water and sediment concentration may not be the maximum values that the canal is expected to convey but those that produce minimum net erosion or deposition during one crop calendar year.

For a given discharge and geometry the actual water depth in a canal under uniform flow condition is decided by the roughness. Roughness keeps on changing throughout the canal operation. The roughness in the bed will change with the change in flow conditions. The roughness in the sides might change due to the growth of vegetation, weathering of canal slopes and periodic maintenance activities. However, the canals have to be designed by taking the average roughness expected during the irrigation season. Modelling provides an option for a precise representation of these changes during the irrigation season that will increase the reliability of predicted morphological changes and help in a better design from sediment transport perspective.

A flow control system is needed in an irrigation scheme to manage the water flows at bifurcations to meet the service criteria and standards regarding flexibility, reliability, equity and adequacy of delivery. A flow is regulated through water level control, discharge control, and/or volume control that make the flow non-uniform. For flows other than the design values, the gates are operated to maintain the set-point and diverting the desired water to the laterals. This will create drawdown or backwater effects and non-equilibrium sediment transport conditions. The canals are designed assuming a steady and uniform flow and an equilibrium sediment transport condition. The sediment transport equations used in the design are not capable of predicting the sediment transport behaviour under non-equilibrium conditions. Sediment transport models provide an option for predicting the sediment transport process in time under changing flow conditions. Hence, a design should be evaluated by using a sediment transport model and necessary changes should be made to reduce the erosion/deposition. Hence, modelling becomes the integral part of the design and it will be helpful to:

- select the design discharge and sediment concentration that will give minimum net erosion/deposition;
- incorporate in the design the changing nature of roughness due to changing hydraulic and management conditions;
- include the effect of water delivery schedules and flow control in the design; and
- prepare and propose the irrigation management plans.

5.5   Conclusions

With a slight change in water flow and sediment properties the sediment movement pattern is affected significantly. Operation and maintenance of the irrigation scheme has major influence in the hydrodynamic behaviour of the canal and hence in the sediment movement also. Analyzing the problem only from hydraulic point of view is not sufficient to solve the problem. An integrated approach that looks into hydraulic as well as management aspect simultaneously is needed to deal the sediment movement problems in irrigation canals. The canal design for sediment transport is, therefore, an iterative process where the starting point should be the preparation of management plans. The design parameters are derived from the management plans and the preliminary hydraulic design of the canal is made. The preliminary design results are then evaluated for the changing water and sediment inflow conditions with the water delivery and operation plans. The results of the evaluation help to fine tune the design parameters and the management plans. The canal dimensions are again calculated using the revised design parameters and the process is continued until the desired level of performance in terms of the sediment movement and water delivery is not obtained.

For the evaluation of the design, a sediment transport model capable of simulating management aspects of irrigation canal is needed.

For the hydraulic calculation an improved design approach has been proposed. The major features can be summarized as:

– *determination of roughness.* The proposed method makes use of the elaborated and more realistically determined roughness value in the design process. The roughness of the section is adjusted as per the hydraulic condition and sediment characteristics. Moreover the influences of the side slopes and the B-h ratio are included while computing the equivalent roughness of the section. This should result in a more accurate prediction of hydraulic and sediment transport characteristics of the canal and hence, a better design.

– *explicit use of sediment parameters.* The explicit use of sediment concentration and size makes the design more flexible as different canals might have to convey sediment loads of different sizes and amounts. This feature gives an option to design and provide an in-canal settling pockets to reduce the sediment size and concentration, if necessary, and design the canal downstream of that point with the new sediment parameters. This will help to design a canal that is easier to operate and reduce the sediment problems.

– *use of adjustment parameter for B-h ratio and side slope.* Since the sediment transport predictors assume the canal to be wide, they do not include the influence of the side walls in shear stress and velocity distribution along the canal cross section. An adjustment parameter has been derived that includes the influence of non-wide canals, sloping side walls and exponent of the velocity in the sediment transport predictor. This adjustment should increase the accuracy of the predictors when they are used in irrigation canals, an environment for which they were not derived.

The possibility of erosion can be checked by comparing the tractive force of the designed canal section. But the design is not based on the erosion control requirements. This is because, this design procedure is proposed for designing a canal that has to carry water with a sediment load. So far as the sediment load in the water is comparable to the design value, then under normal conditions, the energy of flow will be just sufficient to transport the incoming sediment and there will be no erosion of the canal bed. However, if the canal is designed for a certain sediment concentration (say 700 ppm) and due to some reason (construction of a new settling basin or reservoir) the inflow sediment concentration is reduced (say 100 ppm), then there will be possibility of erosion.

The objective of the design should be clear, since designing the canal network that conveys all the sediment down to the field, may keep the canal free of sediment deposition but at the same time the sediment will be accumulated on the agricultural field every year that might affect adversely in soil quality and productivity.

Flexibility of operation and sediment transport aspects restrict each other. A canal without any control can be designed and operated with a higher degree of reliability in terms of sediment transport. Once the flow is controlled the sediment transport pattern of the canal is changed and the designed canal will behave differently. Hence, both flexibility and efficient sediment management are difficult to achieve at the same time. A compromise has to be made and this should be reflected in the design.

In Nepal, the present concept of the design of canal for sediment transport is not in line with the government policy of transferring the management of lower order canals to the farmers. The design is focused to make the canal sediment free, that is, all the sediment should be transported to the lower order canals. The design guidelines provide emphasis in the design of main and secondary canals using sediment transport aspect. The lower order canals are not given importance from sediment transport perspective. The guidelines suggest using simple Manning's equation for the design of these canals. Since, less attention is given to the design of lower order canals from sediment transport aspect; they have more sediment deposition problems. Due to sediment deposition, these canals are difficult to operate and require more maintenance. This is one of the reasons, why the farmers are not interested to take the responsibility of management. Even the systems that have been handed over to the farmers are not operated properly and in some cases the government had to resume its support to the farmers in system operation and maintenance.

The sediment transport problem in the irrigation scheme is not purely a design issue and hence, a designer should have a proper vision and feelings of the operation and management limitations. The vortex tube sediment excluders provided in SMIS had to be closed completely, because the operational limitations of the scheme were not fully analysed while installing them. Similarly, the manager should have the understanding of the design concept that has been used in the scheme. No design can eliminate the sediment transport problem if the system is not operated as per the design assumptions.

6
Model development

6.1   Background

From a sediment transport perspective, the objective of canal design is to ensure that the erosion/deposition in the canal network is a minimum. The sediment load entering into the canal network may be either transported to the field or deposited at specific locations. This objective is quite challenging to meet for canals carrying a significant amount of sediment and passing through alluvial soils, even during steady and uniform flow conditions.

Water demand in an irrigation scheme is not constant throughout the irrigation season, hence, most of the time the canal is carrying a discharge other than the design discharge. Moreover the flow is regulated by water level and discharge regulators to supply water in time, amount and level to the agricultural field. Hence, the flow pattern in the canal becomes non-uniform. Sediment transport is highly dependent on flow parameters, hence, any change in the flow pattern will have a significant effect on the sediment transport capacity of the canal. Hence, with the constant fluctuation in flow pattern deposition and/or erosion in the canal cannot be avoided depending upon the type of flow at that moment. The major concern should be whether the sediment deposited at one stage is eroded after some time or not. If with proper design and operation a balance could be maintained such that after one irrigation season there is no net erosion or deposition then the canal network can be said to be stable.

Sediment transport in an irrigation canal is largely influenced by the water and sediment inflow, irrigation schedules, flow control systems, maintenance type and frequencies of maintenance. As sediment deposition and/or erosion affects directly the capacity of the canal to deliver water in time and amount, the above mentioned factors have direct influence on the performance of an irrigation scheme. To understand the sediment transport process and to incorporate the process in the design, the influence of the above given variables in sediment transport should be understood first. The mathematical model provides an opportunity to study the interaction of different variables in the sediment transport behaviour of a canal. The accuracy with which the prediction can be made when one or more variables are changed depends upon the mathematical representation of the process in the model. The dynamics of sediment movement follow a complex process and a comprehensive analytical description of it is not possible (Raudkivi, 1990). The accuracy of the relationships that describe the sediment transport processes also determines the accuracy of the model.

Focus of many research works is on the sediment movement in river problems and hence, most theories developed have been for wide open channels. They assumed a uniform velocity distribution across the river and the relationships for sediment transport processes were derived for unit width. Although certain similarities exists between rivers and irrigation canals, there are some inherent differences (Table 6.1) that should be taken into consideration while using these theories and relationships for irrigation canals. Most irrigation canals are non-wide (B-h ratio < 8) and trapezoidal in shape (side slope ranging from 1:1 to 1:3) that influences in the velocity distribution pattern across the canal. Hence, the result will not be correct if these aspects are not taken into consideration while computing the sediment transport in irrigation canals.

Most of the available models are for river problems. The modelling of irrigation canals is more demanding due to the presence of flow control and conveyance structures, changing water demands and gate operation to meet the irrigation requirements and periodic maintenance of canals. The concept of the mathematical model SETRIC (SEdiment TRansport in Irrigation Canals) was initiated by Méndez (1998). In subsequent researches, the model was improved and its applicability tested as a tool for the evaluation of irrigation scheme performance in terms of sediment transport under different operation and sediment transport scenarios (Paudel, 2002, Ghimire, 2003, Orellana and Giglio, 2004, Sherpa, 2005).

Table 6.1 General characteristics of water flow and sediment transport in rivers and irrigation canals (Depeweg and Mendéz, 2007).

Water flow and sediment transport		
	Rivers	Irrigation canals
Water profiles	generally without water level control: nearly uniform flow	water level control: gradually varied flow
Froude number	wide range	restricted using drops and energy dissipaters: Fr < 0.4
Discharge	not controlled: increasing in downstream direction	controlled by operation rules: decreasing in downstream direction
Flow control	almost no control structures	regulated flow: discharge, water level regulators
Width-depth ratio	B/h > 15 (wide canals)	B/h < 7- 8
Velocity distribution	nearly uniform velocity distribution in lateral direction	velocity distribution strongly affected by side walls
Alignment	hardly straight, meandered and braided	mostly straight
Topology	convergent	divergent
Lining	alluvial river bed	man-made canals: lining or no lining
Sediment size	wide range of sediment sizes	fine sediment
Size distribution	graded sediment	nearly uniform
Sediment material	river bed	external sources
Transport mode	suspended and bed-load	mainly suspended load
Bed forms	mostly dunes	mostly ripples and mega-ripples
Roughness	skin and form friction	form friction
Concentration	wide range	controlled at headwork

6.2   Water flow

While schematizing the water flow in irrigation canals two aspects have been considered. Firstly the operational aspects where the water flow becomes non-uniform and unsteady due to the changing nature of water demand and operation of gates to meet the water demand and to maintain the water level for a proper irrigation of the fields. Secondly the sediment transport aspect, where the changes in water flow with time and space are much faster than the changes in sediment morphology.

The amount of water flowing in an irrigation canal during an irrigation season and moreover during the lifetime of the canal is not constant. Seasonal changes in crop water requirement, water supply and variation in size and type of the planned cropping pattern are frequent during the lifetime of an irrigation canal. The operation of gates to adjust to the variation in supply is normally gradual to avoid the generation of a wave front. It is common practice to allow sufficient time to move from one steady state to another steady state by operating the gate in multiple stages.

The Froude number in irrigation canals is normally low (< 0.4) to avoid a wavy water surface and large disturbances in bends and transitions (Ranga Raju, 1981). Moreover for a small Froude number (Fr < 0.6 - 0.8), the celerity of the water movement (c_w) is much larger than the celerity of the bed level change (c_s) (De Vries, 1975):

$$|c_w| >> c_s \qquad (6.1)$$

i.e. $c_w \rightarrow \pm\infty$ as compared to c_s, meaning, for the study of sediment movement and morphological change in the bed the changes in the flow from one stage to another are so fast that the bed change during that period can be assumed to be constant (De Vries, 1987).

Hence the flow in the irrigation canals is assumed to be quasi-steady for the study of sediment transport and morphological changes (Figure 6.1), meaning that the changes in unsteady flow conditions are gradual and that the duration of these changes is considerably shorter than the duration of the steady state and time required for any significant change in bed morphology. Under this condition, equation (3.33) becomes:

$$V\frac{\partial h}{\partial x} + h\frac{\partial V}{\partial x} = 0 \qquad (6.2)$$

$$\frac{\partial(Vh)}{\partial x} = 0 \qquad \therefore \quad q = \text{constant} \qquad (6.3)$$

Similarly equation (3.34) becomes:

$$V\frac{\partial V}{\partial x} + g\frac{\partial h}{\partial x} + g\frac{\partial z}{\partial x} = -gS_f \qquad (6.4)$$

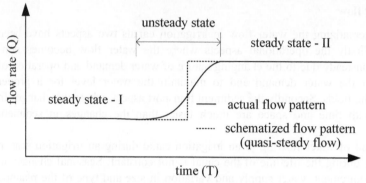

Figure 6.1   Schematization of water flow in irrigation canals.

Assuming a prismatic canal section, small bed slope and uniform velocity distribution, equation (6.4) can be written as:

$$\frac{dh}{dx} = \frac{\left(S_0 - S_f\right)}{\left(1 - Fr^2\right)}$$
(6.5)

with Froude number $$Fr = \sqrt{\frac{Q^2 B_s}{gA^3}}$$ (6.6)

where

A	= area of water flow (m^2)
B_s	= water surface width (m)
g	= acceleration due to gravity (m/s^2)
h	= water depth (m)
Q	= flow rate (m^3/s)
S_0	= bed slope ($-\partial z / \partial x$)
S_f	= friction slope
V	= flow velocity (m/s)
z	= bed level above datum (m)

Equation (6.5) gives the water surface profile of a gradually varied flow. The friction slope (S_f) can be evaluated by any suitable uniform flow formula (Chézy, Manning, Strickler, etc.) assuming that the formula can also be used to compute the friction slope of the gradually varied flow at a given canal section.

Various graphical and numerical methods can be used for the solution of the water surface profile. Here the predictor–corrector method is used, the computation steps of which are summarised in Appendix C (ref. Figure 6.2).

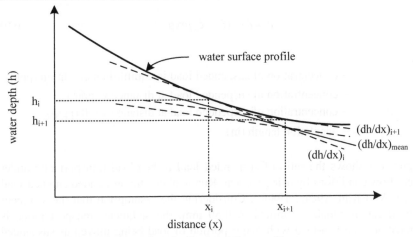

Figure 6.2   Schematization of gradually varied flow profile.

6.3   Sediment movement

For the analysis of the sediment transport process the following basic concepts and assumptions are valid:

−   the sediment can be represented by a single representative size;
−   the size and gradation of sediment remain constant along the whole length of the canal throughout the irrigation season;
−   the canal bed is composed of the same material as that of the inflowing sediment into the canal network.

6.3.1   Galappatti's depth integrated model for non-equilibrium sediment transport

Galappatti (1983) has developed a depth integrated suspended sediment transport model based on an asymptotic solution for the two-dimensional convection equation in the vertical plane. Among the depth integrated model for suspended sediment transport this model has two advantages over others; firstly no empirical relation has been used during the derivation of the model and secondly all possible bed boundary conditions can be used (Wang and Ribberink, 1986). Moreover it includes the boundary condition near the bed, hence an empirical relation for deposition/pick-up rate near the bed is not necessary (Ribberink, 1986). The theoretical background, assumptions and derivation of the model has been given in Appendix D. For steady sediment flow, Galappatti's equation can be written as:

$$\overline{c}_e = \overline{c} + L_A \frac{\partial \overline{c}}{\partial x} \tag{6.7}$$

Integration results:

$$\int_0^x dx = L_A \int_{c_0}^{\overline{c}} \frac{d\overline{c}}{\overline{c}_e - \overline{c}} \tag{6.8}$$

$$\overline{c} = \overline{c}_e - (\overline{c}_e - \overline{c}_0) \exp^{-\frac{x}{L_A}} \tag{6.9}$$

where

$\overline{c}_e$ = concentration of suspended load in equilibrium condition (ppm)

$\overline{c}$ = concentration of suspended load at distance x (ppm)

$\overline{c}_0$ = concentration of suspended load at distance x = 0 (ppm)

L_A = adaptation length (m)

Figure 6.3 shows the ratio of suspended load to bed-load transport rate under different flow conditions by using the Van Rijn methods for suspended and bed-load transport (Van Rijn, 1984a, 1984b). Considering the assumption that the sediment size in irrigation canals is limited to 0.50 mm, the sediment transport mode is suspended and the bed-load with major portion of load being moved as suspended load. Under such conditions the total sediment transport rate predictors will be best suitable. Among the available equilibrium sediment transport predictors the total load predictors are more reliable when tested with the selected data set that are within the range normally found in irrigation canals (Méndez, 1998). Hence, it is imperative that the model for non-equilibrium sediment transport is also for total load transport.

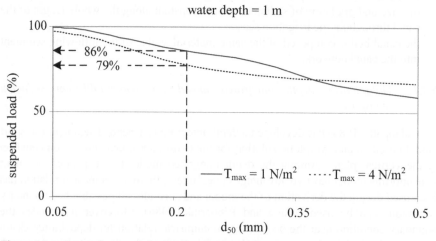

Figure 6.3   Ratio of suspended to total sediment transport rate.

Galappatti's depth integrated model (equation 6.9) is for suspended load only. The assumption that $\omega_s / u_* < 1$, implies that the model is valid for finer sediment ($d_{50} < 0.21$ mm) only (Figure 6.5). For d_{50} less than 0.21 mm, about 80% of the sediment will be transported as suspended load for the flow conditions normally found in irrigation canals (Figure 6.3). Moreover, assuming the adaptation characteristics for suspended sediment to be comparable with the adaptation characteristics of bed movement, the same model is proposed to be used for total load transport. Equation (6.9) can be written in the form of total sediment concentration as:

$$C = C_e - (C_e - C_O)\exp^{-\frac{x}{L_A}}$$ (6.10)

with:

$$L_A = f\left(\frac{u_*}{V}, \frac{w_s}{u_*}, h\right)$$ (6.11)

where

C	= total sediment concentration at distance x (ppm)
C_e	= total sediment concentration in equilibrium condition (ppm)
C_0	= total sediment concentration at distance x = 0 (ppm)
h	= water depth (m)
L_A	= adaptation length (m)
u_*	= shear velocity (m/s)
V	= mean velocity of flow (m/s)
w_s	= fall velocity (m/s)

The concentration will adapt to the equilibrium concentration exponentially and to find the actual concentration at any point x, the variables C_0, L_A, C_e must be known. In the upstream boundary, the initial concentration (C_0) should be known as it cannot be computed from the local conditions. So C_0 in the first reach is the input value which depends upon the source from where the water is flowing into the system. In the subsequent reaches, the computed actual concentration (C) at the end of each upstream reach becomes the initial concentration (C_0) for the downstream reach.

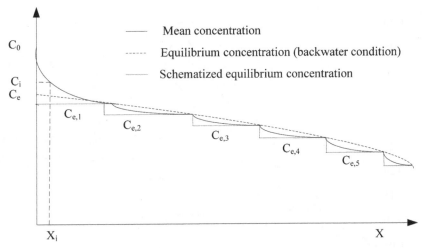

Figure 6.4   Schematization of mean concentration prediction under non-uniform flow condition.

In case, the flow is not uniform the variables, C_e, h, V, u_* and f are not constant but a function of x. If water flow equations are solved separately, then the variables like h and V are known, which can be used to compute Ce, u_* and f. For $i = 1, 2,$

3... n the equilibrium concentration can be computed for the local flow conditions at that reach (Figure 6.4). For the prediction of the actual concentration the hydraulic parameters for that reach can be averaged and the flow pattern can be assumed to be uniform.

Limitation of the model

A detail study of the model was made by Ghimire (2003), Ribberink (1986) and Wang *et al.* (1986). As per their studies, there are certain limitations to this model, most notably are:

- error in the solution increases as the mean concentration moves away from the mean equilibrium concentration:

$$\left|\frac{C_e - C}{C}\right| \ll 1 \qquad (6.12)$$

where, C is local mean concentration and C_e is mean equilibrium concentration. The solution is valid when the deviation of C from C_e is in the range of 0 to 50%;
- the size of the sediment is uniform which can be represented by a single fall velocity;
- ω_s / u_* should be much smaller than unity ($\sim$ 0.3 to 0.4). Figure 6.5 shows the corresponding range of sediment size (d_{50}) for flow conditions normally found in irrigation canals. Considering the maximum permissible tractive force in the bed of canal under normal condition (< 5 N/m^2) the maximum size of sediment for which this model is applicable is 0.18 mm (Ghimire, 2003);
- the maximum fall velocity should be always less than 0.028 m/s.

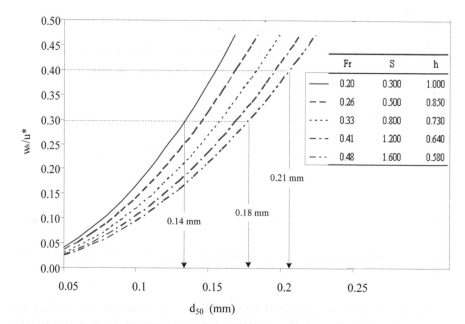

Figure 6.5   Validity range of Galappatti's model under different flow conditions.

6.3.2   Equilibrium sediment transport predictors

After evaluation of various total load predictors, Méndez (1998)has concluded that for the prevailing flow conditions and sediment characteristics (d_{50} = 0.05 to 0.5 mm) in irrigation canals the Ackers and White and Brownlie methods are the best to predict the sediment transport capacity. The Ackers and White method, however, over predicts the transport capacity for sediment size (d_{50}) less than 0.15 mm. The predictors should be used in the similar conditions for which they were derived otherwise the model results might lead to wrong conclusions.

6.3.3   Framework for concentration prediction

The general framework of computation steps used to determine the sediment concentration is given in Figure 6.6. The concentration at the upstream boundary is always known. For each point in the remaining canal the equilibrium sediment carrying capacity of the canal for local flow conditions is computed. Then non-equilibrium sediment concentration profile is solved, depending upon the equilibrium and actual concentration at reference point. The detailed computation steps used to find the non-equilibrium concentration has been summarised in the flow diagram and has been given in Appendix A (Figure A.5).

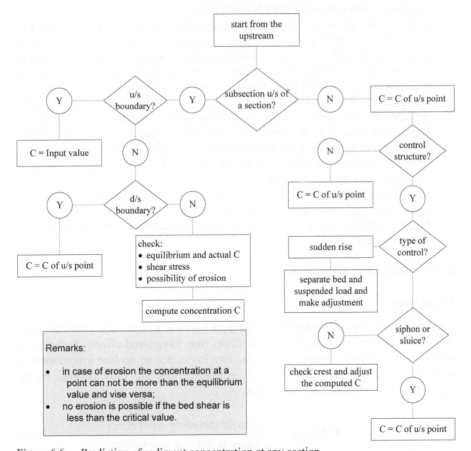

Figure 6.6   Prediction of sediment concentration at any section.

6.4   Roughness prediction

Roughness is a crucial parameter for the solution of the hydraulic and sediment transport relations. The accuracy of its prediction has direct influence in the accuracy of the predicted hydraulic and sediment transport parameters. As discussed in section 4.3 the roughness of a canal section along the perimeter is not constant. In the bed the roughness is determined by the hydraulic condition and sediment characteristics, while in the sides it depends upon the material, condition of canal, weed growth, etc. Moreover the bed width to water depth ratio and the side slopes decide the final equivalent roughness of the section.

Roughness keeps on changing throughout the canal operation. The roughness in the bed will change with the change in flow conditions. The roughness in the sides might change due to the growth of vegetation, weathering of canal slopes and periodic maintenance activities. A precise representation of these changes in the modelling process will increase the reliability of predicted morphological change. The equivalent roughness prediction process involves the following three steps:

– *step I.* The average roughness of the canal before the start of the irrigation season is determined. This roughness indicates the state of the canal before the start of irrigation season. The type and quality of maintenance and age of the canal largely determines the initial roughness. This is the starting point of the roughness computation for both bed and sides;

– *step II.* Determination of the equivalent roughness height in the bed and sides separately. Van Rijn (1984c) method (equations 4.1 to 4.15) is used for the prediction of bed forms and corresponding roughness, since this method provides the best results as compared to other available theories when tested with laboratory and field data (Méndez, 1998). Unlike the beds there will be no change in the slope profile due to deposition/erosion. Any material deposited in the sides will roll down to the bed and no pickup is possible from the sides. This implies that there is no possibility of bed form in the sides. Hence the roughness in the sides is due to the type of material, canal condition and vegetation. At any time the roughness in the side is determined by considering the stage of vegetation, and canal condition;

– *step III.* Computation of the equivalent roughness of the section, considering the aspect ratio and side slope. Equations 4.30 and 4.31 are used as proposed by Méndez (1998) as this procedure yielded the best accuracy when tested with the laboratory data.

6.4.1   Roughness in the sides

Weed effect

For the analysis of the sediment transport process, it is assumed that the canal bed is free of weeds, however, the side slopes may have weed effects. In the cases, where the side slope contains weeds, the roughness due to surface irregularities is neglected. It is assumed that the surface irregularities are fully covered by the vegetation. The weed factor is defined as:

$$\text{weed factor} = \frac{\text{roughness including weed effect}}{\text{roughness without weed effect}} \qquad (6.13)$$

Depending upon the type of weed and relative height of weed compared to the water depth the maximum weed factor is determined. Reference may be made of Table 4.2 that is derived on the basis of suggestions given by Chow (1983). In addition to that information regarding the full growth time of the weed is also needed. From the start to the full growth it is assumed that the influence in the roughness will increase linearly. The maintenance of the canal during the irrigation season may again reduce the effect of vegetation. The duration and frequency of maintenance is also used to determine the weed factor.

Surface irregularities effect

The roughness is only due to the type of material and the surface irregularities present in the sides. The canal condition can be divided into four categories; ideal, good, fair and poor (ref. section 4.5.3). The condition of the canal at the beginning of the season, the expected condition at the end of the season, the condition at the time of maintenance, type and interval of maintenance and the condition after the maintenance are the information needed to predict the roughness in the sides at any time during the irrigation season. It is assumed that the change in canal surface condition with time is linear. The final or intermediate value depending upon whether there will be any maintenance activity or not is to be determined. The condition of the canal may be restored to the original state or improved partially by periodic maintenance during the irrigation season.

Table 6.2   Roughness adjustment factor for different canal conditions.

Canal condition	Multiplication factor
Ideal	1.00
Good	1.27
Fair	1.80
Poor	2.80

Maintenance conditions

The canal condition may be changing during the irrigation season due to the growth of weed, erosion in the toe of the banks, slides, rain-cuts, etc. These aspects would be related to the maintenance condition of the canal to include the effect in the change in roughness. The weed effect is time dependent and can be predicted to some degree, since the type and other properties of vegetation can be easily observed. However, the other factors are difficult to predict. For the analysis of an existing scheme a close observation for a period of two or more irrigation seasons may help to arrive at the most possible scenario. For the design of a new scheme, the information of similar irrigation canals in the vicinity can be used.

The changing state of canal conditions is schematized into three maintenance scenarios; ideal maintenance, adequate maintenance and poor maintenance:

– *ideal maintenance*. It refers to the situation when the canal condition remains the same throughout the irrigation season. The canal may not be in an ideal condition but the canal is maintained at the same status. If there is possibility of weed growth then the weed is removed and any damage in the canal slope is repaired immediately;

– *adequate maintenance*. It refers to the scenario when the maintenance works are planned during the irrigation season to improve the canal conditions. The maintenance may be partial or full. By partial maintenance the canal condition is improved but not to the initial level, while full maintenance means the canal is brought to the initial stage. Maintenance, however, does not include the removal of sediment from the bed and any increase in the canal capacity by widening of the canal bed or increasing of the canal slope. Information like the condition of a canal just before the maintenance, after maintenance and interval of maintenance is needed. The stages of canal before and after maintenance are indicated in terms of initial condition. The stage of canal at any time is determined in terms of the roughness of the side slope assuming a linear relation;

– *poor maintenance*. It refers to the conditions of no intervention whatsoever during the irrigation season. The canal may be deteriorating constantly during the whole season or it may deteriorate for some time and may remain constant after that. The second scenario may be true in case of vegetation, where the full growing time of vegetation is less than the irrigation season. After full growth the effect will be maximum and will remain constant thereafter.

Sequences of calculation roughness and the adjustment of roughness due to weeds and surface irregularities with maintenance and without maintenance scenarios have been given in Appendix A (Figures A.1 to A.3).

6.5   Canal structures and schematization

6.5.1   General

For the conveyance of irrigation water from its source to the farmers field an irrigation network is needed that consists of a large number of appurtenants that can be divided into a conveyance and an operational part. The conveyance part includes the canals and fixed structures like, aqueducts, siphons, bridges, culverts, drops and cascades, flow measuring structures, etc. The operational part includes the structures that divide and control the flow in terms of water level or discharge; they may be fixed without any movable part for the control or they have movable parts like gates.

The design of the conveyance part is normally based on hydraulic and structural requirements, local conditions, available technology and cost. The design of the operational part is based upon additional factors that influence the operation and management e.g. water delivery mode, available manpower, acceptance by the users, ease in operation, transparency, etc. In some designs and layouts different structural arrangements are used within the same scheme. The water delivery and canal operation mode in the Mahakali Irrigation Scheme are different for stage I and stage II and hence the control structures (Pradhan, 1996). Different water delivery modes and structural arrangements can also be found in the Sunsari Morang Irrigation Scheme (FAO, 2006). In some cases, the choice of a structure has been influenced by the background of the designer; e.g. the use of Sharda falls, vertical drop structures and WES-standard falls for the same purpose in Sunsari Morang Irrigation Scheme. The sediment transport aspect is normally not considered in the selection and design of canal structures. Only for the siphons the self-cleaning velocity is the starting point of the hydraulic design. From an operational point of view four major types of structures exist:

- fixed (weirs and orifices);
- on-off (shutter gates);
- adjustable: stepwise (stop logs, modular distributors) or gradually (undershot gates, movable weirs);
- automatic (automatic upstream and downstream water level control structures).

Fixed weirs and orifices require least management input, but they are not flexible in view of water delivery. On-off shutters require more management input and are used for rotational water delivery. Adjustable structures provide flexible water delivery but require a higher operational input. Automatic systems require a smaller number of, but skilled manpower. Hence a balance has to be found in selecting a structure (Horst, 1998).

6.5.2   Hydraulics of flow control structures

The stage-discharge relation of a structure has a direct impact on the overall predictability of a hydrodynamic model. The stage discharge relation and the coefficients vary depending upon the flow stage. From sediment transport modelling aspect, irrigation structures are divided into two categories:
- *structures with sub-critical flow*. Examples of this type are aqueduct, flumed canal section, super-passage, culverts, etc;
- *structures with critical flow*. The flow through these structures is critical/supercritical. The examples of such structures are measuring structures, flow control and division structures, drops, etc. Head loss is more and critical flow may be followed by a hydraulic jump in the downstream for the dissipation of energy.

Structures with sub-critical flow

These structures are designed such that the head loss is a minimum. Smooth transitions are provided to minimize the entry and exit losses. These structures are not used for flow control and depending upon the types these structures may be modelled as:
- a structure with appropriate head loss if the length is short. Sediment deposition, if any, with in the structure is neglected;
- a canal reach with proper geometric properties if the length is significant and in this case the sediment deposition/erosion with in the structure is accounted for in the mass balance of sediment.

Structures with critical flow

The depth discharge relation can be written uniquely for the structures when the flow at control section is critical or modular. However, if the flow is not modular then the upstream water depth is influenced by the downstream water level. It may be due to the sediment deposition, poor maintenance of the canal, or ponding up of the canal in the downstream reach to meet the operational requirements. So a structure with modular flow during some part may turn into drowned flow during other part of the irrigation season. This is common scenario in a contour canal with limited head loss across the structure. In general the depth discharge equation can be written as:

$$Q = \mu CBh\sqrt{2g\Delta h} \qquad\qquad (6.14)$$

where

C	= discharge coefficient
μ	= coefficient for the modular limit
B	= width of control section (m)
Δh	= difference in head (m)
h	= water depth (m)

The values of the variables for different structures and flow conditions are as shown in Tables 6.3 and 6.4.

Table 6.3   Variables for different structures and flow conditions.

Structure	Flow condition	μ	h	Δh
Broad crested	Free	1	$\frac{2}{3}H_1$	$\frac{H_1}{3}$
	Submerged	0.7	H_1	$(H_1 - H_2)$
Flume	Free	1	$\frac{2}{3}H_1$	$\frac{H_1}{3}$
	Submerged	0.8	H_1	$(H_1 - H_2)$
Short crested	Free	1	$\frac{2}{3}H_1$	$\frac{H_1}{3}$
	Submerged	0.2	H_1	$(H_1 - H_2)$
Sharp crested	Free	1	$\frac{2}{3}H_1$	H_1
	Submerged	1	H_1	$(H_1 - H_2)$
Undershot	Free	1	a	$h_1 - C_c a$
	Submerged	1	a	$h_1 - h_2$

where

H_1	= upstream energy head above the crest (m)
H_2	= downstream energy head above the crest (m)
h_1	= upstream water depth (m)
h_2	= downstream water depth (m)
C_c	=contraction coefficient
a	= gate opening (m)

Table 6.4   Range of discharge coefficient (C) for different structures.

Structure	Type of structure/crest	Coefficient
Broad crested weir	Rounded upstream face of crest	Min C = 0.93
		Max C = 1.02
	Sharp upstream face of crest	Min C = 0.848
		Max C = 1.12
Flume	RBC	Min C = 0.93
		Max C = 1.02
Short crested weir	WES weir	C = 0.75
	Sharda type fall (trapezoidal crest)	C = 0.675
	Sharda type fall (rectangular crest)	C = 0.622
Sharp crested weir		Min C = 0.611
		Max C = 1.06
Undershot		Min C = 0.51
		Max C = 0.70

Min = Minimum, Max = Maximum

Effect of sediment deposition

Sediment deposition is not allowed in the bed of flume without a raised crest. Similarly, in case of undershot structures sediment deposition immediately upstream of the gate is not allowed if the gate is flushed with the bed. While, deposition up to the crest level is allowed if a gate is placed over a raised crest. However, it has been assumed that the change in upstream bed level or a relative change in crest height due to a deposition/erosion will have no impact on the stage discharge relation of the gate.

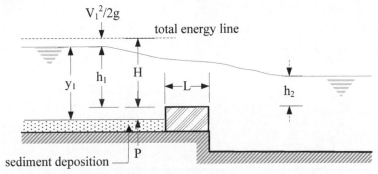

Figure 6.7   Effect of sediment deposition in the over flow structure.

In all other raised crest type of control/measuring overflow structures (Figure 6.7), sediment deposition will change the flow pattern and has been included in the analysis. Depending upon the head discharge relation for a specific flow condition, the upstream water head to maintain the desired flow rate is fixed. If there is deposition in the upstream bed then, the velocity head will increase in the expense of water depth. With the known energy head (H_1) upstream water depth is computed from:

$$H_1 = h_1 + \frac{V_1^2}{2g} = h_1 + \frac{Q^2}{2gB^2 y_1^2} \qquad (6.15)$$

y_1 is the upstream water depth given by:

$$y_1 = h_1 + P \qquad (6.16)$$

where

P　　　　= crest height with reference to upstream bed (m)

　　The deposition of sediment in the upstream bed will reduce the water depth (y_1) and increase the flow velocity. The increase in velocity will increase the sediment transport capacity of the section. Depending upon the incoming sediment load, deposition will continue until the velocity is sufficient enough to convey the incoming load downstream of the structure. The deposition in some cases changes the flow hydraulics and depth discharge relation.

Figure 6.8　　Sediment deposition upstream of flow measuring structure. A flume might have been a better measuring structure than a broad crested weir, in terms of sediment transport efficiency.

　　If the structure is used for flow measurement the gauge reading (depth) may result erroneous result (Figure 6.8). Hence, a proper selection of the structure is necessary if the structure is to be used for a flow measurement or an automatic flow control (without gates) as in a proportional flow divider if the water carries sediment. The steps for the calculation of water depth upstream of a flow control structure used in the model have been summarised in Appendix A (Figure A.4)

6.5.3   Discharge calculation in upstream control

All the irrigation schemes in Nepal have upstream control. Hence in this research only this control option will be studied for sediment transport. Depending upon the water level regulation in the main (on-going or parent) canal and discharge regulation in the off-take, three combinations are possible:
- fixed or passive water level regulator (broad crested weir, Duckbill weir, control notch) in the main canal and a constant head orifice (CHO) with on-off gate in the off-take (Figure 6.9a). This type of arrangement is found in stage II area of SMIS below tertiary canal level;
- fixed regulator in the main canal and gated discharge regulator in the off-take (Figure 6.9b);
- gated water level and discharge regulators (Figure 6.9c). This combination is in all the main and secondary canals of modernized irrigation schemes.

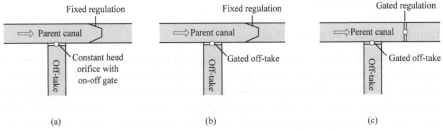

(a)   (b)   (c)

Figure 6.9   Possible combination of water level and discharge regulation in upstream control.

Fixed regulation and CHO off-take

The discharge through the off-take is a function of the upstream head. Since the regulator is fixed, any change in flow rate in the main canal will result in a corresponding change in upstream head. The head and flow rate in the off-take are calculated in the following way:
- the flow rate is computed from the upstream head of the previous time step:

$$Q_{o(i,j)} = f(h_{i,j-1}) \qquad (6.17)$$

where

i	= position subscript
j	= time subscript

- the flow over the regulator ($Q_{2(i,j)}$) is the difference of the incoming flow ($Q_{1(i,j)}$) and the off-take flow:

$$Q_{2(i,j)} = Q_{1(i,j)} - Q_{0(i,j)} \qquad (6.18)$$

- the head in front of the regulator is computed from $Q_{2(i,j)}$ and with the stage discharge relation of the regulator;

– the computations are repeated for the newly found new water depth and iterated till sufficient accuracy is obtained.

Fixed regulation and gated off-take

For the fixed regulation and gated off-take, the discharge is regulated by the gate and measured by a discharge measuring structure downstream of the off-take. Water level regulation is not possible and therefore any change in flow rate in the main canal will result in a change in upstream water level. In this case there are two operation modes possible: the off-take gate is adjusted to release the predetermined (agreed) flow rate or the gate setting is not changed and the flow rate is allowed to fluctuate with the upstream water level. In the first case the discharge through the off-take is known and is deducted from the main canal flow to find the water depth upstream of the regulator. In the second case (gate fixed), the same procedure as in the CHO off-take is used.

Gated regulator and gated off-take

The water level in the combination of gated regulator and gated off-take is regulated and maintained at the set point. Since, the set point is fixed, the upstream water level is known. The opening (flow area) of the undershot or overflow structure can then be computed using the stage discharge relation. Any change in flow rate will result in a change in gate setting. Since the water level is fixed, the flow rate towards the off-take is known. Theoretically the change in flow in the main canal will not affect the flow to the off-take.

6.5.4   General remarks

Automatic flow control in an irrigation scheme with either upstream, downstream or volume control is based on the water level in the canal. Hence the depth discharge relation of the canal is important for a proper functioning of the automatic controls. In canals with sediment problems, the depth discharge relation can not be controlled. The change may arise due to the deposition/erosion or change in roughness during canal operation. Hence, the sediment transport aspect has to be analysed in a proper way before going for the canal automation.

Flow control needs extra care while modelling networks with sediment transport. The hydraulic conditions will keep on changing due to erosion or deposition in the upstream or downstream reaches. Even for the same flow, free flow conditions may convert to submerged conditions or vice versa after some time. Unauthorized obstruction of flow to divert water to both legal and illegal off-takes makes it difficult to schematize (Figure 6.10). In some cases, the manipulations are difficult to trace and go unnoticed throughout the irrigation season. The error introduced in the prediction of the hydraulic condition is again transferred to the sediment transport prediction. The following situations are common:

– if the set point is low and there is deposition in the downstream reach, then a condition may be reached, where the flow is not possible if the set point has to be maintained. This means that a proper stage discharge equation requires a minimum head loss over the structure. If, due to deposition in the downstream reach the water level rises and the available loss is less than required, no water can flow. This aspect needs special attention when modelling the flow in a network with sediment transport.

- if the structures are not properly designed, sediment deposition in the upstream might be possible. A broad crested weir or a sharp crested weir with a raised crest may be changed into a sudden drop. In this case not only the discharge coefficient but also the discharge equation will change;
- in the model it has been assumed that there will be no deposition immediately downstream of the structure, due to the turbulence;
- in a siphon, the upstream water level should be above the invert; otherwise the flow will be a gravity flow as in flumes or open canals. For the same reason the upstream water level (set point) should be more than the sum of the downstream water level and total head loss (Δh). Problems may arise when large discharge fluctuations occur during different periods of the irrigation season;
- in the model it has been assumed that there are no changes in discharge or sediment within the structure. Whatever quantity of water or sediment is going in will come out.

Figure 6.10   Obstruction to raise the water level.

6.6   Morphological change and numerical scheme

The ultimate goal of the sediment transport model is to determine the morphological change that will take place in the canal bed with the given hydraulic, sediment, operation and management variables. The sediment balance equation can be written as:

$$\frac{\partial Q_s}{\partial x} + B(1-p)\frac{\partial z}{\partial t} = 0 \qquad (6.19)$$

The set of water flow and sediment transport equations is solved by an uncoupled formulation. In the uncoupled solution, in one time step Δt, first the water flow equations are solved for velocity, water depth and discharges along all the grid points. In the solution it is assumed that the bed elevation Δz is not changed during this time Δt. The water depths (h) and flow velocities (V) found in the first step are used in the sediment transport to compute the transport rate and the same values are used to compute the bed level change. The same grids are used for the bed level computation that were used for the flow computation.

The uncoupled solution method has its origin in the general concept of explicit finite difference schemes (Figure 6.11). It is assumed that one dependent variable (in this case h) can be computed independently with other dependent variables during one time step. If this is not the case, computational instabilities will appear whatever numerical method is chosen for the solution of equation (6.19), even when water flow equations are unconditionally stable from a strictly numerical viewpoint (Cunge, *et al.*, 1980). These equations are solved alternatively. First the water flow equations are solved (using the predictor-corrector numerical method) and the results of the calculations are used in the next step to solve the sediment transport equation.

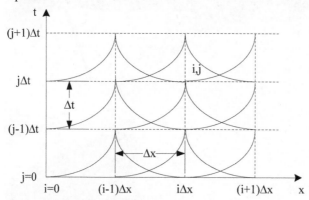

Figure 6.11   Finite difference scheme.

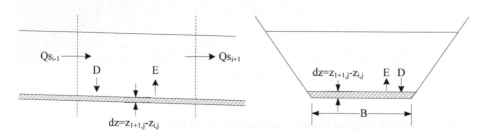

Figure 6.12   Schematization of a canal section for the numerical solution of a bed level change.

Equation (6.19) can be solved by using implicit or explicit numerical schemes. Lax, modified Lax, Leap frog, Lax-Wendroff are some of the explicit methods. The modified Lax method can be expressed as (ref. Figure 6.12):

$$z_{i,j+1} = z_i - \left[\frac{1}{B(1-p)} \frac{\left(Qs_{i+1,j} - Qs_{i-1,j}\right)}{2\Delta x} - \frac{1}{4\Delta t} \left\{ \begin{matrix} \left(\alpha_{i+1,j} + \alpha_{i,j}\right)\left(z_{i+1,j} - z_{i,j}\right) \\ -\left(\alpha_{i,j} + \alpha_{i-1,j}\right)\left(z_{i,j} - z_{i-1,j}\right) \end{matrix} \right\} \right] \Delta t \quad (6.20)$$

This numerical scheme cannot be applied to the downstream and upstream boundaries. An adapted scheme to the downstream boundary is described by:

$$z_{i,j+1} = z_i - \left[\frac{1}{B(1-p)} \frac{\left(Qs_{i,j} - Qs_{i-1,j}\right)}{\Delta x} + \frac{1}{2\Delta t} \left\{ \left(\alpha_{i,j} + \alpha_{i-1,j}\right)\left(z_{i,j} - z_{i-1,j}\right) \right\} \right] \Delta t \quad (6.21)$$

and for the upstream boundary by:

$$z_{i,j+1} = z_i - \left[\frac{1}{B(1-p)} \frac{\left(Qs_{i+1,j} - Qs_{i,j}\right)}{\Delta x} - \frac{1}{2\Delta t} \left\{ \left(\alpha_{i+1,j} + \alpha_{i,j}\right)\left(z_{i+1,j} - z_{i,j}\right) \right\} \right] \Delta t \quad (6.22)$$

where the subscripts i and j mean:

i	= position subscript
j	= time subscript

and

B	= bed width (m)
p	= porosity
Q_s	= sediment discharge (m³/s)
z	= bed level (m)
α	= parameter used for stability and accuracy of the numerical scheme.
Δt	= time step
Δx	= length step (m)

The stability of the scheme is given by (Vreugdenhil, 1989):

$$\sigma^2 \le \alpha \le 1 \quad (6.23)$$

Equation (6.20) is the general form of an explicit scheme and by giving different values to α different schemes can be made. A scheme of the intermediate type can be found if (Vreugdenhil and De Vries, 1967):

$$\alpha = \sigma^2 + \beta \quad (6.24)$$

The accuracy of the modified Lax scheme can be adjusted by means of the parameter β (Abbot and Cunge, 1982). Accuracy of this scheme is increased if $\beta = 0.01$ (Vreugdenhil and Wijbenga, 1982):

$$\alpha \approx \sigma^2 + 0.01 \qquad (6.25)$$

σ is called the Courant number and is described by:

$$\sigma = c\frac{\Delta t}{\Delta x} \leq 1 \qquad (6.26)$$

where c is the celerity of the bed material. Moreover if ψ is a dimensionless transport parameter defined as:

$$\psi = \frac{dq_s / du}{h} \qquad (6.27)$$

Then using the general form of the sediment transport rate (q_s) per unit width for depth averaged velocity (u) and exponent (N):

$$q_s = Mu^N \qquad (6.28)$$

$$\frac{dq_s}{du} = MNu^{N-1} \qquad (6.29)$$

Substituting dq$_s$/du in equation (6.27) results:

$$\psi = N\frac{q_s}{q} \qquad (6.30)$$

Also for smaller Froude number (Fr) < 0.6 to 0.8 the dimensionless transport parameter can be related to celerity (c) as (De Vries, 1993):

$$\phi = \frac{\psi}{1 - Fr^2} \qquad (6.31)$$

where

$$\phi = \frac{c}{u} \qquad (6.32)$$

This gives the celerity of bed movement as:

$$c = Nu\frac{q_s / q}{1 - Fr^2} \qquad (6.33)$$

Combining equations (6.26) and (6.33) gives:

$$\sigma = N\,u\frac{qs / q}{1 - Fr^2}\frac{\Delta t}{\Delta x} \qquad (6.34)$$

For the whole section with mean flow velocity V, the courant number (σ) is given by:

$$\sigma = N\,V\,\frac{Qs/Q}{1 - Fr^2}\,\frac{\Delta t}{\Delta x} \qquad (6.35)$$

where

Fr = Froude number

N = exponent of velocity in the sediment transport equation

Q = water flow rate (m³/s)

Q_s = sediment transport rate (m³/s)

Δt = time interval

ΔX = length step (m)

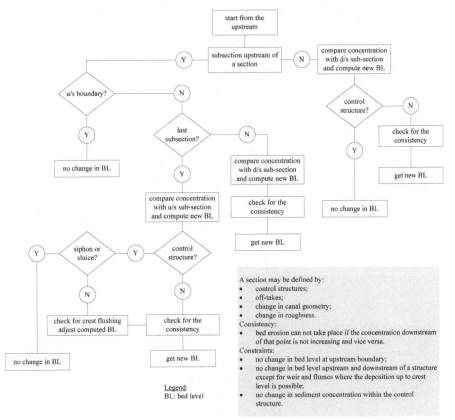

Figure 6.13   Calculation process of morphological change of the bed in SETRIC.

The general procedure used in the SETRIC model is given in Figure 6.13. The detail steps for the solution of sediment balance equation and calculation of bed level change have been summarised in Appendix A (Figure A.6). In summary the computation process of the sediment transport model SETRIC is given in the schematic flow diagram (Figure 6.14).

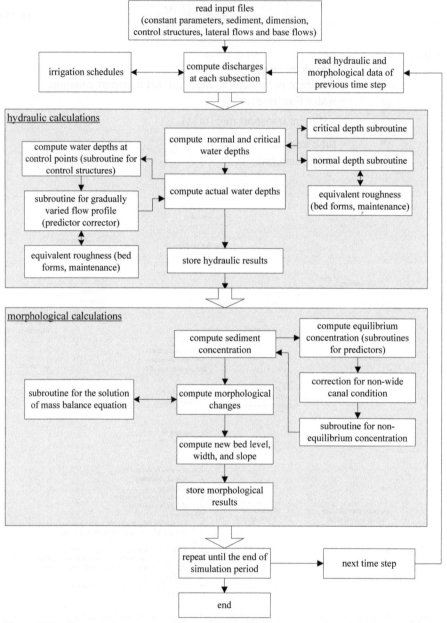

Figure 6.14   Computation process of the model SETRIC.

7
Data collection and scheme analysis

The objective of the field data collection was to gather the information needed to test the assumptions and derived relations for water flow and sediment transport in this research. The sediment transport theories are mostly based on laboratory data where the flow is uniform and the range of hydraulic and sediment characteristics is limited. In practice, one normally encounters a wide range of hydraulic and sediment characteristics in the field and these relations have to be tested before using them for a particular case. The field data will be used to validate the mathematical model SETRIC and the validated model will be used to test the assumptions for and overall performance of the proposed canal design method for sediment transport.

For the field measurements of the sediment transport process, Secondary Canal S9 of Sunsari Morang Irrigation Scheme, Nepal was selected. Since, the objective of field data was to test the design approach for sediment transport; preference was given for a canal that was recently designed and constructed. The Sunsari Morang Irrigation Scheme has faced a serious sedimentation problem in the canal network and a number of remedial measures has also been taken at the headworks. Hence, it is presumed that the designers, during the modernization of the canal, have considered sediment transport aspects in their design. Other considerations of the selection were operation and maintenance conditions, accessibility and working environment in the field.

The field data collection involved two activities, namely, general data collection and field measurement. The general information about the scheme design and management was collected from the design reports, design and construction drawings, and with the water management team and WUA. The information on the water and sediment inflow into the canal, the water delivery schedules (in terms of volume, duration and frequency) to the sub-secondary canals, the gate openings of water level regulators and water levels in the canal, the morphological change after the irrigation season and the maintenance activities and condition of the canal were recorded from field observations. It was observed that the canal operation downstream of the water level regulator at 7,066 m was not regular due to damage of the canal bank by a local drain, hence the measurements up to 7,066 m were made out of the total 13,400 m length of Secondary Canal S9. The field measurements were made during the irrigation of the paddy crop in 2004 and 2005.

The socio-political situation, available equipment, manpower and time had significant effects on the data collection process and the quality of the collected data. Care was taken to maintain the optimum accuracy and uniformity in data collection. For collecting sediment samples and analysis of the collected samples, trained manpower with experience in the related field was involved. As far as possible, the same persons were engaged in the field data collection throughout the season. However, unavailability of standard measuring instruments in case of sediment sampling was a setback in the data collection process.

In this chapter the assumptions and the methods used in collecting data for water, sediment and gate operation are discussed. The measured data is then analysed and the summary is presented. Analysis of the hydraulic design and water management aspects of the system are also presented that is based on the design reports and the observed water management practices being followed in the field.

7.1   Definitions and canal nomenclature

During the modernization, the names of secondary, sub-secondary and newly developed lower order canals have been named in systematic order. Previously, the canals had been named normally after the name of a village in the command area. With the new system, it has become easy to identify the location and the level of the canal. The schematic layout plan and the naming system used in Secondary Canal S9 is shown in Figure 7.1. This naming system, however, is not uniform in the whole area of the Sunsari Morang Irrigation Scheme (SMIS).

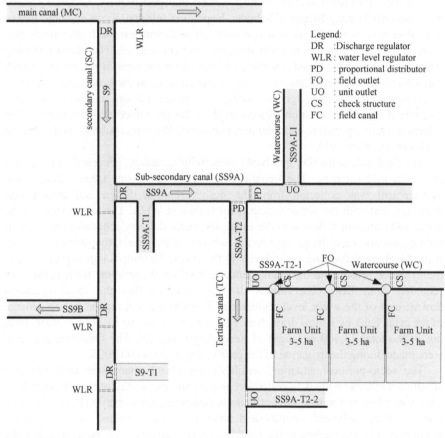

Figure 7.1   Nomenclature and schematic layout of Secondary Canal S9.

The definitions of the terms used in the nomenclature are:

Main Canal (MC)	:	Main Canal of SMIS.
Chatra Main Canal (CMC)	:	Main Canal of SMIS.
CCA	:	Cultivable Command Area.
Secondary Canal (SC)	:	A canal off-taking from and serving at least two sub-secondary canals or tertiary canals.
Sub-secondary Canal (SSC)	:	A canal off-taking from a SC and serving at least two tertiary canals or watercourses.
Tertiary Canal (TC)	:	A canal off-taking from a SC or SSC and serving at least two watercourses.
Watercourse (WC)	:	A canal off-taking from a SC, SSC or TC and serving an average area of 30 ha, called service area.
Field Canal (FC)	:	A canal off-taking from a WC and serving an area of 3-5 ha.
S9	:	ninth secondary canal off-taking from the MC.
SS9C	:	third (A, B, C) sub-secondary canal off-taking from Secondary Canal S9.
S9-T1	:	first tertiary canal directly off-taking from Secondary Canal S9.
SS9C-T1	:	first tertiary canal off-taking from Sub-secondary Canal SS9C.
SS9F-R1	:	first watercourse off-taking from Sub-secondary Canal SS9F to its right side.
SS9F-T1-1	:	first watercourse off-taking from Tertiary Canal SS9F-T1.

7.2   Water requirements and design

7.2.1   Irrigation water requirements

SMIS was originally designed as a supply based scheme with the objective of providing supplementary irrigation water to prevent crop failure (Figure 7.2). During modernization of SMIS, the concept of a demand based allocation was used to compute the irrigation water requirements. Since the available water for the whole command area (CA) of 58,000[1] ha was limited to 45.3 m^3/s the cropping pattern was adopted accordingly (Figure 7.3). However, the introduction of summer paddy was an ambitious concept, since the available water as well as the infrastructure provided did not support for the reliable irrigation of this crop. The available water at the head of Secondary Canal S9 is 5.6 m^3/s for a command area of around 7,900 ha. After taking into account the conveyance and application efficiencies (Table 7.1), the flow rate at field level is sufficient to provide a water gift of 2.9 mm/day. The field application efficiency of 65% seems to be conservative in case of paddy. During the

---

[1] Measured command area is 68,000 ha that includes 10,000 ha sandy area, which is excluded from the water balance calculation (Department of Irrigation, 1987).

first phase of modernization the application efficiency for paddy and other crops was taken as 90% and 70% respectively (Department of Irrigation, 1987). Considering the present assumptions the overall efficiency of the scheme for paddy is 42%.

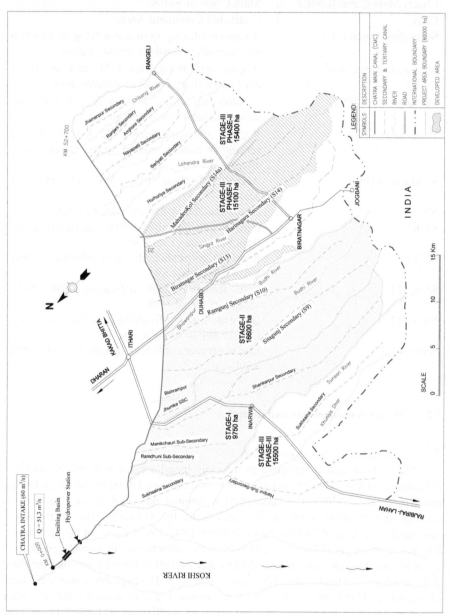

Figure 7.2   Layout map of SMIS.

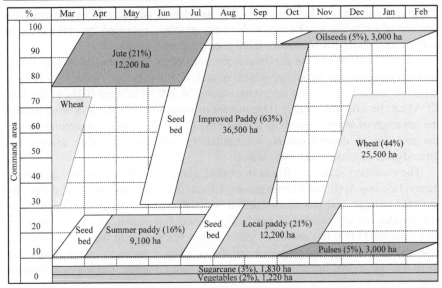

Figure 7.3   Proposed cropping calendar of SMIS (Department of Irrigation, 1987).

The proposed cropping intensity is 180% with paddy as the major crop (Figure 7.3). Paddy has been proposed to be cultivated on 57,800 ha, approximately 55% of the total cropped area (104,550 ha) in one crop calendar. Wheat covers an area of 25,500 ha, which is around 24% of the cropped area in one crop calendar. Hence these two crops (about 79%) decide the net irrigation requirement for most of the irrigation season.

Table 7.1   Water availability at different levels of the irrigation scheme.

System level	Efficiency	Duty for continuous operation
	%	l/s.ha
MC intake		0.78
SC intake (MC off-take)	90	0.71
SSC intake (SC off-take)	92	0.65
WC intake (TC off-take)	92	0.60
FC intake (WC off-take)	85	0.51
Field application	65	0.33

Comparing the capacity of all canals to supply irrigation water at field level (0.33 l/s.ha) operating at full capacity with the net irrigation requirement, the supplied water is not sufficient to meet the irrigation requirements for 7 out of 12 months. Most of the time the available water is not sufficient to irrigate the whole command area, hence the crops have to be grown on a smaller area and most importantly the same pattern needs to be followed throughout the command area. During the month of October the irrigation requirement at field level for local paddy is 7 mm/day (Table 7.2). Local paddy has been planned to be cultivated on an area of 12,200 ha only. With the full supply at the head of the main canal, the demand of

7 mm/day can be met for around 24,000 ha. Hence, particularly in this case, the demand of 7 mm/day can be met without any problem, provided that not more than half the area under any secondary canal is cropped with local paddy.

For the main paddy the maximum demand occurs in the month of July during land preparation and puddling, which is around 8.1 mm/day. With 80% reliable effective rainfall of 5.1 mm/day, only around 36,500 ha of land (63% of the total CCA) can be irrigated reliably (Department of Irrigation, 1995b). It is observed that the coverage of summer paddy is around 90% of the CCA, meaning around 30% of the area has to depend entirely on rainfall or look for alternative sources like groundwater or reuse of drainage water.

The condition is more difficult in case of summer paddy, where the daily water demand during April and May is around 10 mm/day. The value is derived from the monthly average, and if a shorter period is considered, the demand is much higher (15 mm/day towards the end of April). That means that around 20% of the command area (around 11,000 ha) can be irrigated with a full supply in the MC. In the crop calendar only 9,100 ha is proposed and theoretically it seems to be reasonable, but there are operational limitations to supply the required water. The major difficulties are:

– *water diversion problems*. The irrigation scheme receives water from the Koshi River through a side intake. The flow in the river becomes very low during the months of February, March and April. After April the snow melting starts and the river level starts to rise. The crest level of the side intake is 107 m+MSL and the average water level of the river near the intake during the month of April is 107.5 m+MSL. With an available head of 0.5 m above the crest the design discharge of 45.3 m^3/s cannot be diverted. During this season the maximum possible diversion is around 30 m^3/s. That means it is not possible to operate all the secondary canals continuously;

– *operation and management problems*. The designed water delivery schedule is a 1:2 rotation mode[2], hence with the full supply in Secondary Canal S9 only half of the total SSCs can be operated at full capacity at a time. Under continuous operation, only around 40% under each SSC can be supplied with the required demand. Hence the proposed cropped area has to be distributed in all the secondary canals proportionally. But without a full discharge from the intake it is not possible to operate all the secondary canals, hence not possible to follow the cropping calendar.

With the limited available water there is no possibility of selecting crops by the farmers. The coverage of crops like jute and sugarcane, being industrial crops, is going to fluctuate with time. The farmers will switch over to the most profitable crop and hence water stress during the peak demand is possible. This is a common scenario in most of the irrigation schemes in Nepal. The majority of the farmers is small landholders and there is no possibility of commercial farming, hence changes in cropping pattern are frequent, especially in the winter and summer crops.

---

[2] 1:2 rotation mode means the off-takes are divided into 2 groups and one group gets water at a time.

Table 7.2   Irrigation interval for SMIS (Department of Irrigation, 1987).

Month	Main crop	Net irrigation requirement	Allowable soil water depletion	Irrigation interval
		mm/day	Mm	days
Jan	Wheat	2.2	54	24
	Pulses	2.9	60	21
Feb	Wheat	2.6	54	21
Mar	Wheat	1.3	54	41
Apr	Summer paddy	$5.0^{(1)}$	42	8
		$4.5^{(2)}$	$30^{(3)}$	7
May	Summer paddy	$6.2^{(4)\,(1)}$	42	7
		$4.2^{(4)\,(2)}$	$30^{(3)}$	7
Jun	Summer paddy	3.7	42	12
July	Improved paddy	4.2	42	10
Aug	Local paddy	5.0	42	8
Sep	Improved paddy	2.7	42	16
Oct	Local paddy	7.0	42	6
Nov	Local paddy	5.0	42	8
Dec	Pulses	3.0	60	20

(1) Crop water requirement, (2) percolation requirement, (3) permitted depletion of ponded water (4) net after deduction of 1.6 mm effective rainfall.

7.2.2   Water delivery plan

Irrigation interval

The irrigation intervals depend upon the moisture holding capacity of the soil, the allowable soil water depletion, effective root zone for the specific crop and rate of depletion of the moisture. For the soils of SMIS, the available soil water is 120 mm/m and the allowable soil water depletion is 60 mm/m (Department of Irrigation, 1987). The irrigation interval is then computed considering the net crop water requirement at that instant. In a multi-crop scheme like SMIS, different crops will have different irrigation intervals. Considering the major crop of each month, the maximum interval of irrigation is summarised in Table 7.2.

The correct selection of the irrigation interval is difficult in this irrigation scheme, as one farmer may be facing water shortage, while the other may have difficulty to drain the water and protect his field from over irrigation. From a water management perspective, it will be easier, if all the farmers cultivate the same type of crop. That, however, may not be possible, since the available water imposes restriction on the maximum area to be cropped. In Secondary Canal S9, the design irrigation interval is taken as 7 days. This interval satisfies the irrigation demand of the crop in most of the cases.

Duty and unit stream flow

The irrigation supply is rotated over the command area during peak demand in a 1:2 rotational mode, that is, group A for 50% of the command area and group B for the remaining. The unit stream size for the service area in the design is about 1.2 l/s.ha. The service area of a watercourse unit is 28 ha that gives a flow rate (main d'eau) of 32.4 l/s to be managed by a farmer or a group of farmers irrigating around 4 ha for a period of approximately 12 hours per week during peak demand. Thus, there is a 1:7 mode of operation below the level of the watercourse unit. This means that 7 field outlets are connected with a watercourse; each one is being irrigated for 0.5 day at an interval of 7 days.

Scheme operation plan

This irrigation scheme can be classified as a demand based, imposed allocation type. The type and area of the crop is predetermined and the operation plans are prepared assuming that the proposed cropping plans are followed. There is no possibility of change in water allocation, because in the first place the water from the source is not ensured and secondly the infrastructure does not support this kind of flexibility. The method of water acquisition in the main canal, possible water delivery scenarios to the secondary canals and water allocation to the sub-secondary and tertiary canals under peak, off-peak, sufficient water or water shortage conditions are shown in Figure 7.4. Since the diversion of required water to the main canal is not assured, an insufficient supply during peak and even during off-peak periods is possible. Hence, it is important to analyse how the scheme can be operated and what will be the implications on the water management and sediment transport. As shown in the Figure 7.4 there are seven (7) possible operation modes.

Operation mode 1

The system has been designed for the operation mode I. The canal capacity and the control systems have been provided to support this mode of operation. It was assumed that, there would be a variable supply to the main canal, but would be sufficient to meet the demand throughout the irrigation season. The secondary canals during peak demand would be operated continuously and would receive water in proportion to their command area. The command area of the secondary canal has been divided into two groups and each group would receive water in turn. Hence the sub-secondary and tertiary canals are operated in an on/off basis and would receive either full supply or no supply.

The scheme can be operated in this condition for a certain periods of the irrigation season only. Most of the time the discharge in the canal is insufficient to meet the demand from the command area. From water management perspective this mode is the easiest one, since the scheme is designed to run in this mode and the infrastructure has been provided to support the design assumptions. There should be no problem of sediment transport in sub-secondary and lower order canals. The sediment transport capacity of Secondary Canal S9, however, depends upon the arrangements of the off-takes in rotation groups and the degree of non-uniformity they introduce in the canal during operation.

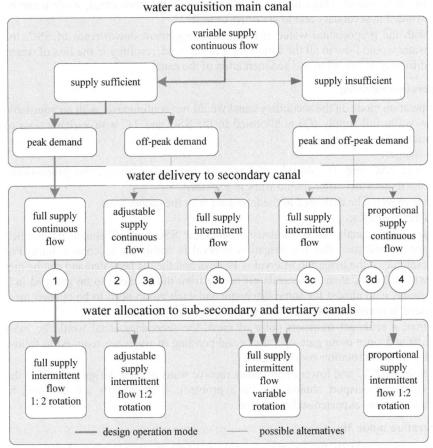

water acquisition main canal

Figure 7.4   Operation modes for supplying water to the secondary canal.

Operation mode 2

Operation mode 2 would be possible when the supply in the main canal is sufficient to meet the off-peak demand. Secondary canals run continuously with an adjusted supply and water allocation to SSC and TC is with the same rotation (T) but adjusted flow rate (Q).

Advantages:

– ease in operation. No possibility of confusion in operation, since the same rotation mode is followed throughout the season;
– possible to manage a large scheme even with poor communication facilities, as farmers already know the timing of their water delivery;

Disadvantages:

– adjustable supply to secondary canal requires frequent setting of the water level and discharge regulators, thus more management cost;
– canals (MC, SC, SSC and TC) flowing with less than the design capacity would have sediment deposition problems, which could increase when the water level

has to be raised. This situation is unavoidable in the main canal, while it can be avoided in secondary and lower order canals;

– with the proportional water distribution arrangement downstream of SSC, the water would flow in all the canals for a lower head, resulting in the loss of water, difficulty in irrigation and sedimentation of the canals.

Operation mode 3a

In operation mode 3a the secondary canal would run continuously with an adjustable flow, while full supply (Q) is allocated to the SSC and TC with variable time of rotation (T).
Advantages:
– since SSC and TC would get water to their full capacity, the proportional distribution thereafter should work as expected;
– water would be delivered to the farm gate with the proper head;
Disadvantages:
– secondary canals run continuously, hence the SSC and TC should be grouped and rotated such that the irrigation interval is within an acceptable limit. For example, if the irrigation interval is 14 days and the off peak demand can be met within 2 days, then the laterals off-taking from the SC need to be grouped in 7 groups with almost the same discharge and each group need to be opened for 2 days. It is not always possible to make such combinations;
– from a sediment transport point of view, the secondary canal would be most affected since more gate operations and ponding of water are required to follow the different rotation combinations;
– since the SSC and lower order canals receive water at their design capacities, the sediment transport should not be a problem, if the canals are designed to transport the expected sediment load.

Operation mode 3b

In operation mode 3b, the secondary canal (SC) runs in rotation but with full supply (Q).
Advantages:
– sediment deposition problems would have to be reduced as the SC runs at full capacity and requires least gate operation;
– the problem of making rotation combinations of SSC to match the new demand and irrigation interval would not arise;
– less management input as compared to operation mode 3a;
Disadvantages:
– managing the water may be a problem for the farmers in multi crop farming;
– adopting a specific rotation of the secondary canals is difficult as the duration and magnitude of water shortage in the main canal can not be predicted.

Operation mode 3c

Operation mode 3c may be adopted when the available water in the main canal is insufficient. Secondary canals run at full capacity for limited periods. So from a hydraulic and management perspective this mode is similar to the mode 3b.

Operation mode 3d

In this case the available water in the main canal is distributed proportionally to the secondary canals. Depending upon the available water in the SC, the SSCs would be grouped and operated in rotation modes. This mode would be similar to 3a, but there are extra difficulties to operate the scheme due to the uncertainty at the source. It would be difficult to make rotation plans and to inform the farmers accordingly. Without a strong communication system, this mode of operation would be difficult to implement.

Operation mode 4

Operation mode 4 is similar to operation mode 2 with an intermittent water allocation to SSC and TC. The difference is that in this case the supplied water would have insufficient head and quantity.

Existing operation mode

For the analysis of the water delivery practices being followed in Secondary Canal S9, the problem can be divided into two phases. In the first phase, immediately after the modernization, the system was operated adhering to the design mode. In the first place the assumption that despite the uncertainty in the water diversion the available water in the main canal would be sufficient to meet the demand was not fulfilled. As a result the secondary canals including Secondary Canal S9 received a continuous but changing discharge (Q) for most of the time. Since, no specific plans were given in the design for the rearrangement of the sub-secondary canals off-taking from Secondary Canal S9 during less than design discharge the water management agency followed the design delivery schedule even for the less than design discharge. Thus, the sub-secondary and tertiary canals were run in low heads due to the less than the design discharge. The design water distribution structure in the sub-secondary canal is a proportional divider while that in the tertiary and lower order canal is a unit outlet. The proportional divider could not increase the head of the flow to the tertiary canal. Since, the unit outlets (orifice type) provided in the tertiary canal are less sensitive to the upstream head, the outlets near the head diverted more water than their share leaving less water to the tail enders. Hence, the farmers from the tail end started to block the flow in the sub-secondary and tertiary canals to increase the discharge and head. This activity reduced the sediment transport capacity of the lower order canals and the sediment started to deposit in these canals. Maintenance activities of these canals are not properly done mainly due to:
- the farmers in the upper reach were reluctant to desilt the canal, since the desilting would lower the water level and they would face difficulty in diverting water;
- the operation and maintenance of these canals was the responsibility of WUA and they had shortage of fund to clean the canals to the design state.

Thus, the deposited volume in the sub-secondary and lower order canals kept on increasing every year resulting in the reduction of the canal capacity.

While in the second phase, since, the sub-secondary canals could not carry the design discharge due to improper maintenance, the design delivery schedule could not be followed. Since, no specific water delivery mode was followed, people started to ask for opening their off-take more frequently or for a longer duration. This lead to chaos in the water delivery plan and more sediment deposition in the canals. The farmers have abandoned the water distribution and outlet structures provided in the canal and started to divert water from other places.

As will be discussed in chapter 8, the sub-secondary and tertiary canals were not designed based on sediment transport consideration and the deposition pattern was random. Some canals were functioning well while some were running at less than 50% of the design capacity. Under such condition, the only choice left with the water management agency was to run more sub-secondary canals at a time with less discharge. Hence, the system has now entered into a vicious cycle of deterioration. Large investment (similar to rehabilitation work) is needed to return the system in its design stage.

7.2.3   Canal alignment, control and design

Alignment

The main canal of SMIS is a contour canal, except for the idle length, where it passes through the deep cutting area before reaching the command area. The slope is gentle and there are hardly any drop structures except for the head losses in the water level regulators. The secondary canals follow the ridges and have a number of drops of varying size. Normally natural drains separate the command area of two secondary canals. With the secondary canals essentially along the ridge, the sub-secondary canals are normally aligned along the contour and tertiary canals along the ridges depending upon the topography.

During modernization, the existing infrastructure of main and secondary canals was upgraded to meet the new plans and the canal alignment was not changed. However, for fixing the alignment of the lower order canals the following criteria and considerations were adopted:
- the alignment of the sub-secondary and tertiary canals was planned taking into account the topographic conditions, existing irrigation and drainage network and the boundary of village development committee and ward[3]. This was done to assist the farmers to organize themselves for the operation and maintenance and to reduce conflicts in water sharing;
- the service area of one WC is about 28 to 30 ha. The location of the last field outlet was fixed considering the distance of the farthest field plot, and the area served by it is about 1/7th of the total service area i.e., 4 ha each.

Flow control

From a flow control perspective, SMIS can be divided into two levels; one that is being managed by the irrigation agency (government) and the second one being managed by the WUA. In level one the upstream control is provided, with undershot water level and discharge regulators. The vertical slide gates are manually operated by screw and handle (Figure 7.5). The discharge regulators are provided with the

---

[3] the local administrative boundary known as Village Development Committee (VDC) covers an area of about 800 to 900 ha. There are 9 sub-administrative divisions of a VDC, known as ward (numbered from 1 to 9 and about 100 ha).

broad crested weir at the downstream of the off-take for discharge measurement (Figure 7.6).

Figure 7.5   Water level regulator in Secondary Canal S9.

In the second level, below the head of a sub-secondary canal, proportional flow control is used (Figure 7.7), which is also known as a "structured design" (Easter, *et al.*, 1998). Since there is no gated control in the second level below the sub-secondary canals, farmers cannot ask for water, unless all the farmers under one sub-secondary have the same crop and demand. The overall control mechanism and water delivery policy that are possible in Secondary Canal S9 are:

– intake gates are provided at the head of the secondary canal, sub-secondary canals and tertiary canals (off-taking directly from the secondary canal) for control and diversion of water to Secondary Canal S9 and from S9 to the lower order canals (Figure 7.6). These gates are controlled by the agency responsible for irrigation in SMIS;

– the design mode of operation was 1:2 for the peak demand and it was suggested to follow the same operation during the off-peak demand. During peak demand water is released to nearly half the command area for 3.5 consecutive days. Hence some unsteady flow conditions prevail in the canal network before a new steady state is realized;

– secondary canals receive irrigation water continuously. Water level regulators are provided along the secondary canal to ensure the required supply to the off-takes during off-peak flows;

- for diverting water from sub-secondary to lower order canals, proportional fixed crest weirs in the parent and in the off-take have been provided. This arrangement also supplies water to those canals that have no demand and the water is lost;
- water delivery to the field canals is by a fixed orifice. The level of the orifice is preset to draw the supply as per command area below that point. Since the cropped area and accordingly demand might change, the fixed flow rate cannot fulfil the water demand. That is one reason, farmers in most cases have either changed the setting of the unit outlets or are using other means to divert water;
- the flow rate in the watercourse is equal to the unit flow times the area to be served (main d'eau) and the area is between 18 – 30 ha (average 24 ha), thus giving a flow rate of 22 – 36 l/s (average 24 l/s).

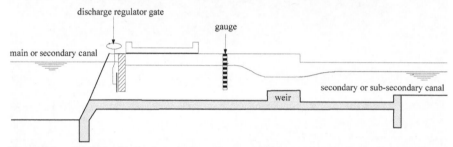

Figure 7.6   Typical off-take for upstream flow control (secondary and sub-secondary canal off-takes).

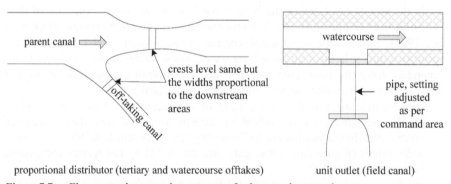

Figure 7.7   Flow control system downstream of sub-secondary canal.

Canal design

All the canals of SMIS including Secondary Canal S9 are unlined and trapezoidal in shape with side slopes normally 1V: 1.5H, except for some critical places, where the canal is lined. The canals are designed for uniform flow. Truly uniform flow seldom occurs in nature, however, this is a logical starting point. The roughness is estimated based on the type of canal and the expected maintenance condition during operation. In other words, the size of an irrigation canal is based on estimated roughness, which is also practically valid for gradually varied flow.

In view of the sediment problems prevailing in SMIS, the sediment transport aspect would have to be included in the design of the canals system. Initially Secondary Canal S9 was designed using Lacey's regime approach (equations 5.1 to 5.5) and the same method was followed during modernization (Department of Irrigation, 1987). However, the sub-secondary, tertiary and lower order canals were designed using Manning's uniform flow equation together with a relation of the B-h ratio (Equation (7.1)). The designed mean velocity was then checked with the Kennedy's critical velocity (V_c).

$$h = 0.5\sqrt{B} \tag{7.1}$$

$$V_c = 0.54mh^{0.64} \tag{7.2}$$

where
B = bed width (m)
h = water depth (m)
m = critical velocity ratio
V_c = critical velocity (m/s)

The critical velocity ratio is a function of sediment size and concentration and for SMIS this ratio was taken as 0.84.

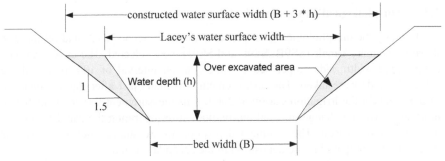

Figure 7.8   Typical canal section and top width as per Lacey's regime equation.

The regime canals as predicted by Lacey's equations would have steeper side slopes (Figure 7.8). From slope stability considerations, the adopted side slope in the scheme (during modernization) is 1V:1.5H. This means, the constructed canal section is larger than the designed section. The actual water depth would be less than the computed one. On average, for a given water depth, the difference in discharge predicted by Lacey's equation and that measured in the field is around 14% (Mishra, 2004).

The flexibility to modify the longitudinal slope is normally restricted due to the presence of drop structures, siphons, aqueducts, and the off-take points during modernization. In Secondary Canal S9 the bed slope of the canal has not been modified, which is steeper than that given by Lacey's regime slope. During the design the bed width, water depth and flow velocity were computed using the regime equations, while the bed slope was based on the field condition. Principally, this is not the correct approach, since the water depth, bed width and flow velocity predicted by Lacey's regime equations are for the regime slope only. The use of regime equations for a canal system that have control structures at every 1 km and drop structures at every 500 m is conceptually not appropriate. There would be less possibility for the canal to adjust itself to the regime slope and width.

Sub-secondary and lower level canals

All the sub-secondary canals including their lower order canals have been designed by Manning's equation. The estimated Manning's roughness for the canals is presented in Table 7.3. No consideration for sediment transport requirement was made for these canals.

Table 7.3   Manning's roughness coefficient (Department of Irrigation, 1987).

Canal	Manning's roughness coefficient $(s/m^{1/3})$
Earthen secondary canal	0.0225
Earthen sub-secondary canal	0.025
Earthen tertiary canal	0.025
Earthen watercourse	0.03

7.3   Morphological and geometrical data

To study the morphological changes, in addition to Secondary Canal S9 three of its sub-secondary canals (SS9B, SS9F and SS9J) were selected. The cross sectional details (bed width, side slope, bed elevation) at an interval of 100 m for all the selected canals were taken. The measurement was made just after the maintenance of the canal for the irrigation season of 2004. This measurement served as the bench mark and gave the change in canal morphology after operating the canal for one or more irrigation season. The sediment deposition in the sub-secondary canals was quite high as compared to the deposition in the secondary canal.

Bed morphology in 2004

In 2004, a comparison of the two sets of measurements showed that there was an average deposition of 8.5 cm in Secondary Canal S9. The deposition was, however, not uniformly distributed and it was more near and upstream of the control structures. As expected the deposition was less near the upper reach of the canal where the water depth and flow velocity were higher. In the first 7.1 km the average deposition was 6.5 cm while in the last 6.3 km it was around 12.6 cm.

Bed morphology in 2005

Since the sediment deposition in 2004 along Secondary Canal S9 was not significant (average depth 8.5 cm), sediment was not removed from the bed. The bed level of the secondary canal was measured before the irrigation season of 2005. Comparison of the two measurements showed the redistribution of deposited material during the winter irrigation of 2005. The canal carried relatively clear water during winter, hence the sediment from the upper reach of the canal was eroded and transported to the lower reaches. Since the discharge during the winter irrigation was small, a significant change in the bed morphology was not possible.

After the irrigation season of 2005, the average deposition in the canal bed up to 7.1 km was 12.5 cm. This amount was significantly higher (around 92%) that that of the 2004 season. The reason could be the increase in sediment concentration and duration of irrigation as compared to 2004.

Geometrical data

The geometric and hydraulic information was collected from design reports and as built drawings. The data was verified by field measurements during the survey work. The secondary canal was found to be maintained to its design geometric shape, however, the quality of maintenance was not uniform throughout the length. Table 7.4 shows the geometric and hydraulic data of Secondary Canal S9.

Table 7.4   Design hydraulic and geometric data of Secondary Canal S9 (Department of Irrigation, 1987).

Reach	Kilo-metrage	Hydraulic conditions			Dimensions of canal			
		Discharge	Bed slope	Velocity	Water depth	Bed width	Free board	Canal depth
	km	Q	S_0	V	h	B	FB	(h+FB)
	km+m	m^3/s	m/km	m/s	m	m	m	m
1	0+000	5.60	0.233	0.64	1.13	6.9	0.67	1.8
2	0+398	4.75	0.242	0.63	1.07	6.3	0.63	1.7
3	4+610	4.25	0.248	0.62	1.03	6.0	0.67	1.7
4	6+694	3.25	0.261	0.59	0.94	5.2	0.66	1.6
5	9+586	2.80	0.270	0.57	0.89	4.8	0.61	1.5
6	11+363	2.30	0.280	0.56	0.84	4.4	0.56	1.4
7	12+323	1.80	0.293	0.53	0.78	3.9	0.52	1.3
8	14+340	1.80	0.293	0.53	0.78	3.9	0.52	1.3

7.4   Water flow data

The flow into Secondary Canal S9 and outflow to the sub-secondary canals were noted throughout the irrigation seasons of 2004 and 2005. Water flow to the secondary and sub-secondary canals is controlled by gated discharge regulators. A broad crested measuring structure has been provided downstream of the discharge regulator (Figure 7.9). The gauge records upstream of the measuring weirs were noted to determine the flow rate to the secondary and sub-secondary canals. In addition to that the water levels upstream of the gates and the gate openings were also recorded for cross checking purposes.

There are seven water level regulators in Secondary Canal S9. The gate openings and the water levels upstream of each regulator were recorded during the whole irrigation season. The records showed that no specific set point was maintained at any water level regulator. The gate operator operated the water level regulator and discharge regulator randomly to deliver the targeted amount of water in the lateral.

Figure 7.9   Broad crested weir near the intake of Secondary Canal S9 canal for flow measurement.

Gauge readings once a day (6 AM) in all the laterals and three times a day (6 AM, 12 noon and 6 PM) at the head of Secondary Canal S9 were taken. The inflow rate into the canal was found to be fairly constant as the gate operator constantly adjusted the discharge regulator to maintain the flow. The sum of outflows from the secondary canal was generally less than the inflow at the head of the secondary canal. The difference between outflow and inflow was around 5% and the discrepancies could be due to one or combinations of the following:
–   seepage loss along the canal that has not been included in the computations;
–   illegal outlets that draw a significant amount of water which were not measured;
–   it takes more than 6 hours to stabilise the flow in Secondary Canal S9 once some disturbance is introduced. So when the inflow is changed at the head the effect could be seen near the tail after only 6 hours. Since the laterals were not running as per design they were opened and closed randomly thus frequently creating unsteadiness in the system.

The flow measurements showed that the inflow hydrograph of Secondary Canal S9 was quite different from the expected pattern (Figure 7.10). In 2004, around 67% of the total irrigation period the inflow was either equal or more than the design inflow (5.6 m^3/s), while in 2005, it was around 47%. On a volumetric basis the total amount of water delivered is approximately equal to the designed total (approximately 101% in 2004 and 96% in 2005) considering the design discharge for the canal opening period only. This comparison is on the assumption that the canal was operated for the design discharge throughout the irrigation period. From water requirement consideration, however, the total volume supplied was quite high which was around 125% in 2004 and 160% in 2005.

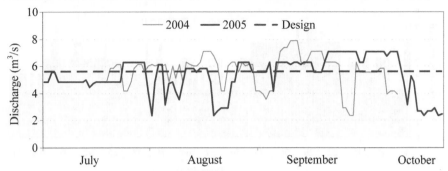

Figure 7.10   Water inflow to Secondary Canal S9 canal for paddy season of 2004 and 2005.

The water depth near the water level regulator was also recorded. This parameter would indicate the flow pattern in a reach for the given inflow. The sediment carrying capacity of the flow and erosion/deposition rate depends upon the flow pattern (uniform or non-uniform flow).

7.4.1   *Rotation schedule*

As per design, the off-takes from Secondary Canal S9 have been grouped in two rotation groups A and B as shown in Figure 7.14. During peak demand there would be a rotation of 3.5 days on and 3.5 days off. Out of the total 14 off-takes (10 sub-secondary canals and 4 tertiary canals), 8 are in group A and 6 are in group B. In actual practice the rotation schedule was hardly followed during the irrigation season. As many as 13 off-takes out of 14 were running at the same time for a certain period (Figure 7.11), while sometimes only 3 out of 14 were running.

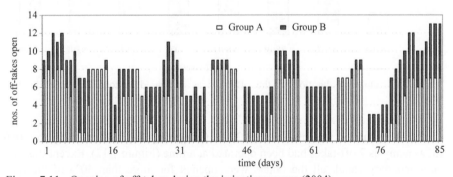

Figure 7.11   Opening of off-takes during the irrigation season (2004).

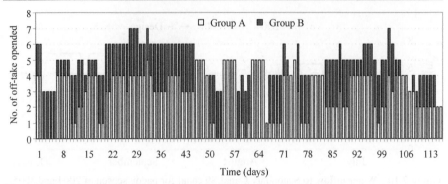

Figure 7.12   Opening of off-takes during the irrigation season (2005).

To run more off-takes than designed, requires either more discharge in the secondary canal or less than the design discharge to the sub-secondary canals. As can be seen in Figure 7.10, the discharge into the secondary canal was more than the design value for more than 67% of the time in 2004. Moreover, each off-take was receiving less than the design discharge. This operation was contrary to the design concept and the flow control structures provided to regulate the water delivery.

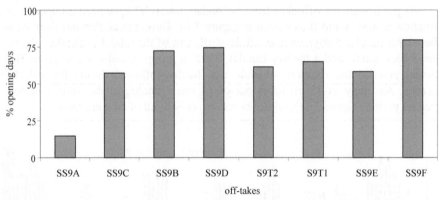

Figure 7.13   Opening duration of the off-takes as compared to total irrigation duration (2005).

Similar situation was observed in 2005. Comparing the 8 off-takes up to 7,066 m, as many as 7 off-takes had been operated at a time (Figure 7.12). Except the S9A sub-secondary canal, all the others were opened for more than 50% of the time (Figure 7.13).

Figure 7.14   Rotation schedule of off-takes from Secondary Canal S9.

7.4.2   Calibration of broad crested weir

The broad crested weir near the head of Secondary Canal S9 was calibrated and compared with the Winflume (Wahl, et al., 2001) results. As can be seen in Figure 7.15 the measured value is comparable with the model result. Hence the flow equation predicted by the model has been used for the derivation of the discharge.

$$Q = 12.4 \, h^{1.57} \tag{7.3}$$

where

h   = head over the crest (m)
Q   = discharge (m³/s)

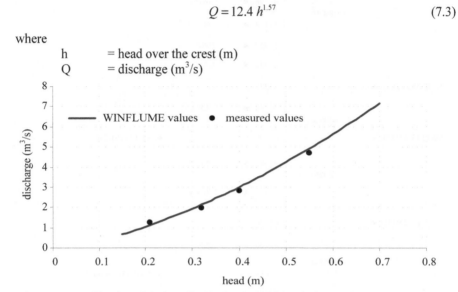

Figure 7.15   Calibration of the broad crested weir at the head of Secondary Canal S9.

7.5   Sediment data

Inflow of sediment into the system was measured near the head of Secondary Canal S9. The measurements were taken daily during the irrigation season of 2004 and 2005. The daily variation of concentration is presented in Figures 7.16 and 7.17. During laboratory analysis two fractions were separated namely the fraction that was greater than 63 μm and a fraction finer than 63 μm. The two fractions were added to obtain the total suspended load.

Sediment concentration

Sediment concentrations were measured using dip samples. This is the simplest way of taking a sample of suspended sediment. A bucket or bottle is dipped into the canal, preferably at a point where it will be well mixed, such as downstream of a weir or drop. The sediment contained in a measured volume of water is filtered, dried and weighed. This gives the concentration of sediment and when combined with the flow rate it gives the sediment discharge. This method generally gives concentrations that are about 25% lower than results obtained from more sophisticated techniques (Rooseboom and Annandale, 1981). The location for taking the samples was just downstream of the discharge regulator where a strong hydraulic jump was present due to a drop in bed level by around 70 cm. At this location the flow is well mixed, and the dip samples can be expected to yield the suspended load concentration that are close to the average concentration. Most of the coarser fraction of the sediment was separated in the pre-settling and settling basin near the intake of the main canal. The intake of the Secondary Canal S9 being at 24.8 km of the main canal, the sediment size was mostly fine sand.

To find the difference in the measurements, some dip and pump samples were taken simultaneously. The bed load part could not be measured due to the unavailability of bed load sampler. Hence, theoretical approach as suggested by Colby (1957) has been used. The total load has been computed by adding bed load to the measured suspended load. The calculation steps have been summarised in Annex B. The results (Table 7.5) show that dip samples underestimated the mean total sediment concentration by an average factor of 0.92.

Table 7.5   Measured pump and dip samples in Secondary Canal S9 (2004).

Date	Dip samples (ppm)	Pump samples (ppm)
16-Aug-04	760	800
17-Aug-04	810	960
18-Aug-04	930	1,120
19-Aug-04	1,160	970
20-Aug-04	1,040	940
21-Aug-04	1,160	1,460
22-Aug-04	1,010	1,360
23-Aug-04	690	640
Average	945	1,030

Considering the sand fraction only (size > 63 μm), the dip samples underestimated the pump samples by around 65% (35% less). HR Wallingford (1988) collected some data from dip and pump sampling downstream of a vortex tube in the SMIS for comparison purposes and found that the concentrations measured by dip sampling were around 30% less than those from pump sampling.

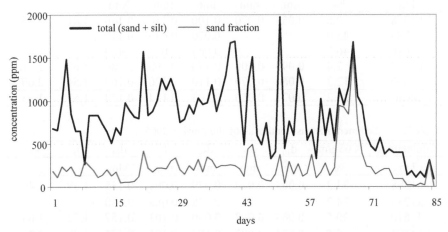

Figure 7.16   Sediment concentration (total) at the head of Secondary Canal S9 during the paddy irrigation season (26 July to 23 October, 2004).

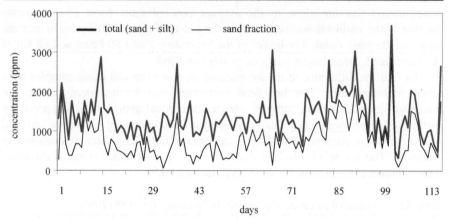

Figure 7.17   Sediment concentration (total) at the head of Secondary Canal S9 during the paddy irrigation season (8 July to 1 November, 2005).

A comparison between the sediment concentration at the head of Secondary Canal S9 shows that the average concentration in 2005 is higher than in 2004. This indicates a reduction in the efficiency of the settling basin. If remedial measures are not taken, then the sediment problem in the scheme will increase.

Sediment size distribution

After the irrigation season some bed material samples were taken along the canal. The deposited material was mostly sand and silt. No sorting of material in downstream direction could be observed. Table 7.6 and Table 7.7 show the sediment size and distribution along the canal in 2004 and 2005 respectively.

Table 7.6   Sediment size distribution along the canal in 2004.

Kilometrage	< 63 μm	d_{10}	d_{16}	d_{50}	d_{60}	d_{84}	σ_g	C_u
km	%	mm	mm	mm	mm	Mm		
0+398	9.8	0.063	0.070	0.100	0.110	0.140	1.41	1.75
3+286	45.0			0.070	0.080	0.110		
4+610	46.7			0.065	0.071	0.100		
6+694	2.4	0.090	0.100	0.170	0.190	0.280	1.67	2.11
7+066	12.3	0.060	0.070	0.110	0.120	0.160	1.51	2.00
Mean	23.2	0.071	0.080	0.103	0.114	0.158	1.53	1.95

Table 7.7   Sediment size distribution along the canal in 2005.

Kilometrage	< 63 μm	d_{10}	d_{16}	d_{50}	d_{60}	d_{84}	σ_g	C_u
km	%	mm	mm	mm	mm	Mm		
0+398	33.2	0.063	0.070	0.079	0.094	0.135	1.39	1.49
3+286	34.2			0.078	0.088	0.130		
4+610	30.5	0.063	0.068	0.090	0.103	0.157	1.52	1.64
6+694	6.5	0.071	0.083	0.170	0.195	0.275	1.82	2.75
7+066	12.7	0.063	0.070	0.100	0.120	0.170	1.56	1.90
Mean	23.4	0.070	0.070	0.100	0.120	0.170	1.57	1.94

where
$$d_n \qquad = \text{n\% of the sample finer than the given value in mm}$$

geometric standard deviation $\qquad \sigma_g = \sqrt{\dfrac{d_{84}}{d_{16}}}$ $\qquad\qquad\qquad$ (7.4)

uniformity coefficient $\qquad\qquad C_u = \dfrac{d_{60}}{d_{10}}$ $\qquad\qquad\qquad$ (7.5)

The analysis showed that the samples contained around 23% of material finer than 63 μm. The median size of the sediment sample was found to be 0.102 mm. The uniformity coefficient is smaller than 2 hence, one representative value can be used for the modelling of the sediment transport (Halcrow, 2003). The grain size analysis result of the sediment samples have been given in Appendix B.

7.6   Conclusions

SMIS was originally designed as supply based system without an assured water diversion provision at the head. After the improvements of the headwork during the rehabilitation, the water diversion from the river has been improved but still it is not assured. Hence, the concept of the design to convert it into a demand based scheme is a difficult proposition to meet. Hence reliability, adequacy and timeliness of irrigation supply mainly depends upon the rainfall in the catchment (to raise the water level in the river) and in the command area (to supplement the crop water requirement).

With the given water diversion and delivery infrastructure, the overall objective of the irrigation scheme would have to be to distribute the available water equitably to the crops in the entire command area keeping in mind that crop stress is unavoidable, even during the wet season. The water delivery plan and cropping schedule are so rigid that there exists very less possibility of changes. That is the reason why with each stage of modernization the cropping schedule is revised. Farmers by default assume that the scheme is a protective one only and there will be water stress if there is no rainfall as expected. Thus, a rigid cropping calendar and water delivery schedule, has to be followed assuming that the Irrigation Management Agency possesses enough knowledge about the crop, market, soil type, climate, cultural practices and water availability. The findings of Secondary Canal S9 on the design and management can be summarized as:
- the sediment transport aspect is considered in the secondary canal only (using Lacey's equations). In sub-secondary and tertiary canals, the sediment transport aspects have not been considered at all;
- the infrastructure provided in the distribution network (below sub-secondary canal) works properly only for the full supply discharge. For the irrigation scheme with such an un-assured water diversion provision, the system is difficult to operate in a proper way;

- sediment transport is a major problem for sub-secondary and tertiary canals. The rate of sediment deposition is different from canal to canal and from reach to reach with in the same canal. Hence, in the system, some farmers have to invest more for sediment removal than the others. This has made the system operation more difficult;
- the water operation plans are not being followed. A canal carrying sediment can not be operated and maintained unless proper water delivery modes (from sediment transport and water requirement aspects) are established and followed strictly;
- due to the some socio-political situation during the data collection strict and regular monitoring of the data collection activity could not be made. Some of the observers might have taken that as an advantage and entered some of the readings in the record books without actually visiting the site. This, however, could not be verified.

8
Application of improved design approach

8.1   General

The proposed improvement in the design of irrigation canals for sediment transport will be evaluated and used in this section. The results will be compared with the existing canal design that is mostly used in the design of irrigation canals in Nepal. The Sunsari Morang Irrigation Scheme (SMIS) is a rehabilitation type of scheme and the rehabilitation work is continuing in different stages. The analysis will help to identify the problems associated with the present design and management practices and take corrective measures to rectify them.

The sediment transport process in an irrigation canal is influenced by the operation and maintenance of the irrigation scheme. The canal can be designed to pass an incoming sediment load under the assumption of uniform flow and an equilibrium transport condition. Once, the conditions deviate from the design value the flow velocity and thus the sediment transport capacity will be changed in time and space. The operation of the irrigation scheme will introduce variations in the design values and assumptions in time and space. Besides, the hydraulic calculations are based on the assumed maintenance condition of the canal. If the canal is not maintained to the design condition, the flow parameters will change. Such changes in the design parameters and the design assumptions introduce the change in sediment transport process. The deviation of the design parameters and assumptions from the design condition is a continuous process during the irrigation season or during the life time of a canal. Hence, the sediment transport aspect of an irrigation canal would have to be analysed considering changing scenarios in time and space.

The mathematical model provides an option to analyse the sediment transport process in an irrigation canal for changing flow and sediment parameters with time and space. Hence, the model SETRIC as described in chapter 6 will be used to simulate the effect on the sediment transport process due to the variation in water and sediment inflow, introduction of water delivery plans, introduction of flow controls and the maintenance plans. The model will be first evaluated comparing its predictability on hydraulic and sediment transport with the existing models, field conditions and measured data of Secondary Canal S9 of SMIS. Then the evaluated model will be used to simulate the different scenarios. The proposed improvements in the determination of the hydraulic roughness, the selection of the sediment transport predictors and the design process will be evaluated. The effect of water delivery plans, control structures and operation of gates on the sediment transport will be evaluated.

In this section, three terms are frequently used namely, normal depth, actual depth and design depth or set point. The normal depth is the flow depth in the canal for a uniform flow condition, while the actual depth is the depth for which the flow may or may not be uniform depending upon the structural control at the downstream end. The design depth or the set point is the target depth that should be met for the diversion of desired water to a off-take.

8.2   Evaluation of the model SETRIC

8.2.1   Model verification

As discussed in chapters 5 and 6, there are two levels of computational environments in SETRIC; firstly the hydraulic calculations are done and then the results of the hydraulic calculations are used for the morphological calculation (Figure 6.14). The hydraulic computation in SETRIC is based on quasi-steady flow conditions, hence it should be comparable with the result of a hydrodynamic model under steady flow conditions. Numerical accuracy of the hydraulic calculations of SETRIC will be tested with DUFLOW, while the sediment transport prediction will be compared with SOBEK-RIVER.

Hydraulic performance

The hydraulic performance of the model has been compared with the results of the DUFLOW model. Hydraulic conditions in SETRIC have been assumed to be quasi-steady, hence the results of the DUFLOW model for steady state should be comparable to the SETRIC results. For this purpose a canal system with the following geometric and hydraulic parameters was used (Table 8.1):
- total length of canal   = 5 km
- constant inflow rate at the head   = 6.0 m^3/s
- constant sediment inflow rate   = 10 ppm
- size of sediment (d_{50})   = 0.1 mm
- Chézy's coefficient   = 40
- maintenance condition of the canal   = ideal, i.e. there will be no change in roughness due to change in canal condition with time.

Table 8.1   Geometric data of the canal.

	length (km)	bed slope (1/1,000)	bed width (m)	side slope (V:H)
reach 1	3.0	0.5	5.0	1:1
reach 2	2.0	0.25	5.0	1:1

However, the two models have some differences; most importantly in the prediction of the roughness and the influence of the morphological change on the hydraulic parameters. For comparison of the hydraulic results the following adjustments were made:
- *roughness*. In case of SETRIC the roughness of the canal section is not constant and it changes with bed width to water depth ratio and the relative roughness in the bed and sides. Roughness in the bed will change with the bed forms and in the sides due to the vegetation and maintenance plan of the canal. While in case of DUFLOW, the roughness is constant. To control the roughness in the bed the canal has been assumed to be non-erodible and the sediment inflow rate was kept small such that there was no deposition. Similarly to control the roughness in the sides the maintenance condition has been taken to be ideal. This assumption would ensure a constant roughness in the canal;

- *morphological change.* Since the hydraulic conditions in SETRIC are influenced by the erosion or deposition of sediment, the effect was controlled by assuming a non-erodible canal carrying very low sediment load.

A weir was placed at 3 km and its width was adjusted to get a backwater flow profile (M1) in reach 1. Similarly, water depth at the downstream boundary was fixed such that the flow profile in reach 2 became a drawdown (M2) type.

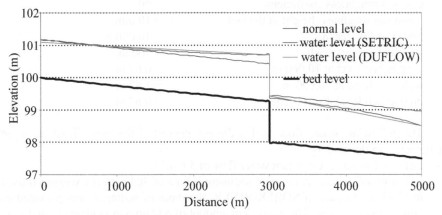

Figure 8.1   Water level predicted by two models.

The hydraulic calculation results by both the models are presented in Figure 8.1. The water surface profiles predicted by the two models are slightly different, because DUFLOW computed the flow depths at two nodes (node distance was 1 km) and the profile between the nodes is assumed to be linear, while SETRIC computed the water depth at every 100 m. In another example, a canal reach of 1 km length was taken and the water level at every 100 m was computed using the two models and the results are shown in Figure 8.2. It shows that the difference (Figure 8.1) is due to the different locations of the computation points in SETRIC and DUFLOW. Hence, the hydraulic computation modules of SETRIC are satisfactory.

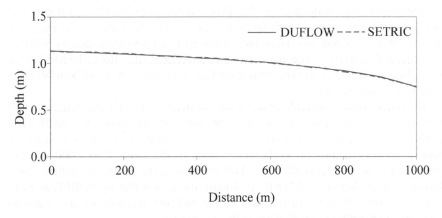

Figure 8.2   Water depth predicted by the two models for the same computational nodes.

Sediment transport prediction

The morphological prediction of SETRIC has been compared with SOBEK-RIVER, which is a morphological 1-D model. For the comparison the following hydraulic and geometric information were used:

- bed width   = 6.9 m
- side slope   = 1:1.5
- longitudinal slope   = 0.233 (m per 1,000 m)
- Chézy's roughness coefficient   = 50
- minimum roughness height in the bed   = 10 mm
- length of first reach   = 10,000 m
- length of second reach   = 500 m
- drop in bed level at 10,000 m   = 0.67 m
- width of flume at 10,000 m   = 3.62 m
- drop in bed level d/s of flume   = 0.67 m

The following water flow and sediment characteristics were used as input parameters in the models:

- *water inflow rate*. A constant water flow of 5.6 m^3/s;
- *sediment inflow rate*. A constant sediment inflow of 500 ppm (by weight) is used in SETRIC. In case of SOBEK-RIVER, the input of sediment was provided as volume per unit time. The equivalent amount of 500 ppm in volume per unit time becomes 0.00106 m^3/s (for a discharge of 5.6 m^3/s and sediment with a specific weight of 2.65 kg/m^3);
- *representative sediment size*. The median sediment diameter d_{50} was taken as 0.1 mm.

The boundary conditions applied were:

- *hydraulic boundary*. A constant inflow rate (Q) as the upstream boundary and the normal water depth as the downstream boundary;
- *morphological boundary*. A constant sediment inflow at the upstream boundary and a constant bed level at the downstream boundary.

SETRIC calculates the equivalent roughness considering the effects of bed width to water depth ratio, roughness in the sides and roughness in the bed due to bed forms. The effect of weed growth on the roughness was eliminated by assuming the ideal maintenance condition. Thus, the equivalent roughness computed by SETRIC was independent of time. The equivalent roughness computed from SETRIC for the given hydraulic and sediment characteristics was used in SOBEK-RIVER to simulate the same scenario.

The model was simulated for a period of 90 days by using the Engelund and Hansen total load predictor as sediment transport predictor for equilibrium conditions. The results by the two models are presented in Figure 8.3. Both the models predicted deposition in the upper reach of the canal (0 to 5,000 m). The sediment transport trends predicted by both the models are similar, while the volume deposited as predicted by SETRIC is slightly more than that by SOBEK-RIVER. There are some inherent differences in the solution methods of the sediment transport and morphological changes in the two models:

– *effective width.* In case of SOBEK-RIVER the sediment transport (total) is given by (Delft Hydraulics and Ministry of Transport Public Works and Water Management, 1994b):

$$Q_s = q_s W_s \quad and \quad W_s = \alpha_s T \tag{8.1}$$

where

Q_s = total sediment transport rate (m^3/s)

q_s = sediment transport rate per unit width (m^2/s)

W_s = effective width (m)

α_s = reduction factor ($0 < \alpha < 1$)

T = water surface width (m)

In SETRIC the effective width is the mean of bed width and water surface width.

– *sediment transport process.* In SOBEK-RIVER the sediment transport rates or capacities at any two points are estimated using sediment-transport formulae. Then the cross-section change between the two sections are computed using the gradients in the sediment transport rates (Sloff, 2006). That is sediment transport is influenced by the local conditions only. While SETRIC uses Galappatti's model for predicting sediment concentration under non-equilibrium conditions. That is the sediment transport process at any section is influenced by the hydraulic and sediment concentration conditions of the upstream section;

– *solution of mass balance equation.* For the solution of the mass balance equation, SETRIC uses the modified Lax scheme for the solution while SOBEK-RIVER uses Lax-Wendroff type of scheme.

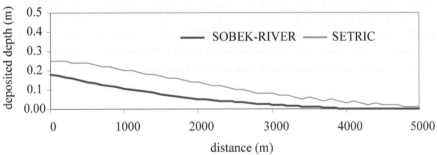

Figure 8.3   Deposition depth predicted by two models.

Considering the calculating environment of the two models, the morphological results predicted by SETRIC are in line with the results of SOBEK-RIVER.

It can be concluded that the results of SETRIC for water flow and sediment transport are comparable with the models that have been well tested and have been widely used for practical and research purposes. Hence, the model SETRIC can be used for the analysis of sediment transport in irrigation canals and networks.

Practical aspects of morphological change

Any man-made canal or natural channel that is in equilibrium with a certain water and sediment discharge, will continue to be in equilibrium as long as one of the two (water or sediment flow rate) is not changed. If, however, the water flow or sediment parameters are changed and maintained during a sufficiently long duration, the canal would adapt itself to the new condition by adjusting either the bed width or the bed slope. This aspect has been tested with the model. There is no provision of adjusting the bed width in the model; hence it is assumed that the new equilibrium condition will be achieved by adjusting the bed slope only. A canal with the following parameters was used:

- length of canal (L) $= 10,000$ m
- bed width (B) $= 6.8$ m
- side slope (m) $= 1V : 1.5H$
- bed slope (S_0) $= 0.23$ m per 1,000 m
- design discharge (Q) $= 4.5$ m^3/s
- normal water depth (h) $= 0.91$ m
- sediment size (d_{50}) $= 0.10$ mm
- Chézy's roughness coefficient $= 46.5$ m$^{1/2}$/s
- sediment transport capacity of the
  canal under equilibrium condition $= 220$ ppm
- sediment transport predictor $=$ Brownlie

An undershot type gate was placed at the downstream boundary to maintain a constant water depth of 0.91 m. The model was run for the design discharge and sediment concentration. Since, the design condition prevailed in the canal; the carrying capacity (equilibrium sediment concentration) is equal to the actual sediment concentration (Figure 8.4).

Then the incoming sediment concentration was increased from 220 ppm to 300 ppm without changing the hydraulic parameters. Since, the incoming sediment load was more than the carrying capacity of the canal deposition should take place. The model was run continuously for a period of 3 years and the results are shown in Figure 8.5. The results are in line with the sedimentation process that is generally observed under such hydro-morphological conditions.

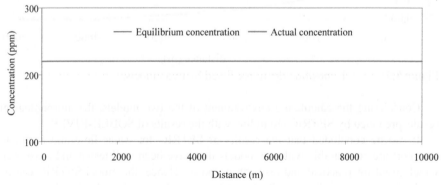

Figure 8.4   Equilibrium and actual concentration under design condition.

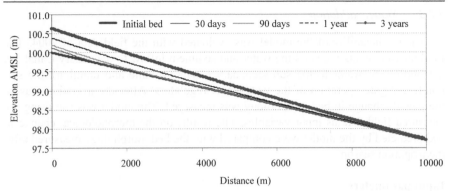

Figure 8.5   Change in canal bed level with the increased sediment concentration.

Since the incoming sediment load was more than the carrying capacity of the canal, the excess sediment load would start depositing near the head reach. The deposition would make the bed slope steeper and thereby increasing the transport capacity locally. Hence, lesser sediment would be deposited at the section under consideration and more sediment would be passed to the subsequent section downstream where then the excess amount would be deposited. Thus, two processes occurred simultaneously, firstly increment in the deposition depth at any section and secondly extension of the deposition in downstream direction. The model predicted that the deposition extended to the upper 2,000 m of the canal within 30 days and around 5,000 m in 90 days. In one year the deposition took place in the whole reach, while the depth in the upper reach kept on increasing. After three years the bed level of the canal near the head was raised by 0.63 m that gave a new bed slope of 0.29 m per 1,000 m. The equilibrium sediment transport capacity of the canal for this slope was around 300 ppm which was also the incoming sediment load.

8.2.2   Model calibration

Roughness was taken as the calibration parameter. In the model, roughnesses in the sides and in the bed are determined separately and then the equivalent roughness of the section is computed (ref. section 4.3 and 6.4). Since roughness in the bed depends upon the hydraulic and sediment parameters, roughness of the sides has to be adjusted to arrive at the measured mean or equivalent roughness of the section. The roughness of the canal was measured three times in one irrigation season (start, mid and end). It was found that the roughness remained practically the same throughout the season. The roughness was slightly high at the beginning of the irrigation season, decreased during the mid season and remained constant thereafter till the end of the season.

The average Chézy's roughness coefficient (C) derived from the measured hydraulic and geometric parameters of Secondary Canal S9 during the irrigation season was 44.0 m$^{1/2}$/s in 2004 and 44.6 m$^{1/2}$/s in 2005. The equivalent value in terms of Manning's roughness was 0.0224.

8.2.3   Model validation

The predictability of the model was evaluated with the field data of 2004. The field measurements included the water and sediment inflow to Secondary Canal S9, the outflow of water to the sub-secondary canals (volume and duration), the set-points and gate operation of the water level regulators. After the irrigation season, the overall change in bed morphology was measured. The model was run using the measured field data as input variables. The results on the morphological change in bed predicted by the model were compared with the bed morphology measured after the irrigation season.

Input parameters

Figures 7.10 and 7.16 show the water and sediment inflows into the canal during the paddy season of 2004. The Secondary Canal S9 received water for a period of 85 days. In case of sediment concentration, only the sand fraction (size > 0.63 µm) of the total concentration was used. Although, the water flow, gate operation and delivery schedule up to 7,066 m was recorded, only the first reach from 0 m to 398 m was used for validation purpose. This was done to reduce the effects of other variables, like illegal water withdrawal, interference to the flows by farmers, performance of the hydraulic structures, etc. that could not be incorporated in the model but still would have influence on the sediment transport process.

In the reach from 0 m to 398 m, there are 3 off-takes, 2 off-takes (for SS9A and SS9C) at 50 m and 1 off-take (for SS9B) at 356 m from the head and a water level regulator are located at 398 m. The daily measurement of water delivery to the off-takes (volume and duration) and the water level upstream of the water level regulator were used. The operation schedules and discharge to the sub-secondary canals up to the water level regulator at km 7+066 are given in Appendix B. Other general parameters used for the modelling are:

- representative sediment size (d_{50})   = 0.102 mm
- sediment transport predictor   = Brownlie
- canal length   = 7,066 m
- simulation period   = 85 days (2004)

Model calibration

The model was calibrated using the roughness as the calibration parameter for uniform flow conditions. The average Chézy's roughness of the canal for uniform flow conditions was derived from field measurements and was found to be 44.0 $m^{1/2}$/s. During the calibration a Chézy's roughness value of 38 for the side slope of the canal for the design discharge of 5.6 m^3/s gave an equivalent roughness of 44.0 $m^{1/2}$/s. Since, there was no weed growth on the sides during the irrigation season, the effect due to weed was neglected by assuming an ideal maintenance condition.

Prediction of roughness

The roughness predicted by the model in time and space has been presented in Figures 8.6 and 8.7. Figure 8.6 shows the roughness along the canal after 12 days. The predicted roughness along the canal is more or less constant, but it is slightly more than the roughness for uniform flow conditions. Figure 8.7 shows the variation of roughness with time.

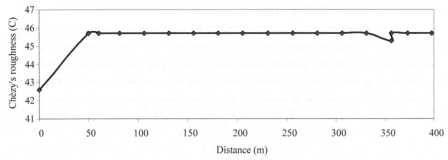

Figure 8.6   Predicted equivalent Chézy's roughness coefficient in Secondary Canal S9 on 12th day of canal operation.

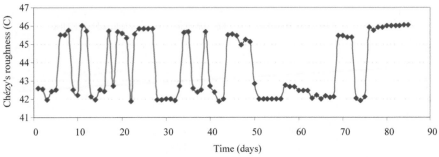

Figure 8.7   Predicted equivalent Chézy's roughness coefficient at 180 m from the head of Secondary Canal S9.

The following comments can be made regarding the change in Chézy's roughness value in time and space:

- *the change in flow due to opening and closing of the off-takes at 50 m.* The operation of the off-take for Sub-secondary Canal SS9C is more pronounced compared to that for SS9A as the design capacity of SS9C is around 3.5 times more than SS9A. The change in flow alters the water depth and hence the B-h ratio. The change in B-h ratio changes the influence of bed and side on the equivalent roughness;

- *Chézy's roughness formula.* The Chézy's roughness coefficient depends upon the hydraulic depth and any change in water depth is reflected in the change in roughness;

- *critical shear stress and bed forms.* The roughness in the bed is more when bed forms are created, that is when the shear stress at the bed is more than the critical value needed to move the sediment. This is also the condition for entrainment of deposited sediment to the flow if the capacity of the flowing water is more than the existing sediment load. Hence, the lower Chézy's value indicates the condition of high shear stress and possibility of sediment entrainment to the flow or erosion of deposited volume. While, a higher Chézy's value indicates the condition of low shear stress and no movement of bed materials. The condition also indicates the possibility of sediment deposition.

Prediction of sediment deposition

The deposition along the canal predicted by the model and that measured after the irrigation season is shown in Figure 8.8. The deposition trend shown by the model is in line with the measurements in the field. The difference in predicted and measured values near the upstream and downstream end is due to the presence of off-take structures at 50 m and 356 m where the sediment deposition was not allowed in the model to maintain the intended flow into the off-takes. The model has predicted a total deposition of around 79 m^3 which is around 67% of the total measured volume within this reach. Although, the objective of the validation was not to verify the model quantitatively, the result is an indication that the proposed improvements in the sediment transport predictors by incorporating the effects of side slope and B-h ratio and the method used to predict the bed roughness and to calculate the equivalent roughness have improved the predictability of the model.

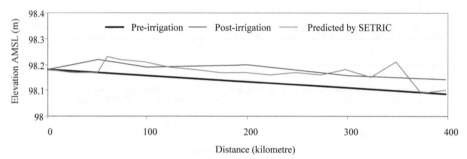

Figure 8.8   Comparison of deposition trend predicted by the model and measured after the irrigation season of 2004.

The objective of this research is to understand the effects of design, operation and maintenance on the sediment transport process more in qualitative terms than in quantitative and to use this model for the comparative evaluation of the existing canal design and management practices with the proposed improvements in the design and management of irrigation schemes in view of sediment transport. Despite the difficulties of schematizing the real flow situation, incorporating some of the management related issues as well as the actual condition of the irrigation network, the modelling results are in line with the observed sediment transport process and hence the model SETRIC has been used for further analysis.

8.2.4   Predictability of equilibrium predictors

The same model setup as that used in sub-section 8.2.3 for the irrigation season of 2004 was used to evaluate the predictability of the equilibrium sediment transport predictors. Three total load predictors, i.e., Engelund and Hansen (EH), Brownlie (BR) and Ackers-White (AW) were evaluated in view of their predictability. The predicted accuracy of the three predictors using the field data of 2004 and 2005 for 7,066 m length of canal is shown in Figure 8.9.

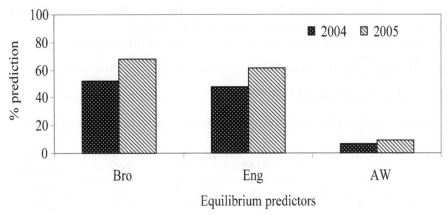

Figure 8.9   Prediction accuracy of the predictors.

The predicted accuracy is measured as a percentage prediction, which is the ratio of the predicted volume to the measured volume of sediment deposited in the canal. Figure 8.9 shows that the Brownlie predictor's prediction accuracy was around 60% (52% for 2004 and 68% for 2005, while the accuracy of Engelund and Hansen was around 50% (48% for 2004 and 61% for 2005). The prediction accuracy of Ackers and White method was low at around 9% (7% for 2004 and 9% for 2005). Hence, for the conditions prevailing in Secondary Canal S9, Brownlie and Engelund and Hansen predictors are more suitable. In the subsequent analysis these two predictors has been used.

8.3   Canal design aspects

Secondary Canal S9 was designed using Lacey's equations that do not require specific information on sediment size and concentration. Due to the differences in the design calculations and the constructed canal (refer 7.2.3 canal design), the actual water depth has been found to be less than the design value. If Secondary Canal S9 is allowed to run without control, the off-takes will not draw their share of water as the crest level of the off-takes has been set considering the design water level. Hence, the flow has to be controlled by using water level regulators to maintain the target water level even for the design discharge. For the analysis of the design from a sediment transport point of view, the canal was allowed to run normally without any control and it was assumed that the off-takes would be able to draw their share of water.

For making a comparison three different cases were considered:
– case-I: canal with the existing design and construction;
– case-II: canal with the existing design but adjusted bed width to get the target water level;
– case-III: canal designed by using the improved design method.

Based on the design discharge and canal geometry, Secondary Canal S9 up to 7.066 km can be divided into 4 parts. The design geometry (case-I) with the following hydraulic and sediment characteristics was evaluated for the sediment transport capacity under equilibrium conditions. The evaluation was made by using Brownlie's predictor and the result is presented in Table 8.2.

- mean sediment size (d_{50})   = 0.10 mm
- relative density   = 2.65
- side slope of canal   = 1:1.5
- roughness (Manning's) coefficient   = 0.022

Table 8.2   Sediment transport capacity of Secondary Canal S9 for normal water depth.

Length		Discharge	Bed width	Bed slope	Normal depth	Equilibrium sediment concentration
From	To					
m	m	m³/s	m	m/1,000 m	m	ppm
0	398	5.60	6.9	0.233	1.06	227
398	4,600	4.75	6.3	0.242	0.99	228
4,600	6700	4.25	6.0	0.248	0.95	229
6,700	7,066	3.25	5.2	0.261	0.85	228

The equilibrium transport capacity shown in Table 8.2 is obtained when the canal was allowed to flow freely without any obstruction. As the design water depth was more than the normal depth by around 8%, the actual sediment carrying capacity of the canal was less when the design depth was maintained (Table 8.3). The design water depth in this case was obtained by increasing the roughness of the section, so that the flow in the canal was uniform.

Table 8.3   Sediment transport capacity of Secondary Canal S9 for design water depth.

Length		Discharge	Bed width	Bed slope	Design depth	Equilibrium sediment concentration
From	To					
m	m	m³/s	m	m/1,000 m	m	ppm
0	398	5.60	6.9	0.233	1.13	201
398	4,600	4.75	6.3	0.242	1.07	198
4,600	6,700	4.25	6.0	0.248	1.03	188
6,700	7,066	3.25	5.2	0.261	0.94	180

In case-II, the existing bed slope, side slope and water depth (target/design) were used to compute the bed width by using Manning's equation. The resulting section was then evaluated for the sediment transport capacity under equilibrium conditions using Brownlie's predictor (Table 8.4). It was found that the adjustment in the width improved the transport capacity by around 25 ppm (around 12%) as compared to normal flow condition and by around 50 ppm (around 25%) as compared to design water depth condition. It would not only reduce the necessity of gate operation but also the construction (small canal section) as well as maintenance cost.

Secondary Canal S9 was a rehabilitation type of scheme; hence flexibility in the design was limited. However, by making minor adjustments in the bed slope and water depths, the sediment transport capacity of the canal could be increased.

Table 8.4   Sediment transport capacity of redesigned Secondary Canal S9 under equilibrium condition.

Length		Discharge	Bed width	Bed slope	Water depth	Equilibrium sediment concentration
From	To					
m	m	m³/s	m	m/1,000 m	m	ppm
0	398	5.60	6.1	0.233	1.13	254
398	4,600	4.75	5.6	0.242	1.07	256
4,600	6,700	4.25	5.2	0.248	1.03	260
6,700	7,066	3.25	4.5	0.261	0.94	259

SMIS has a settling basin at the head of the main canal. As per the design, the basin would limit the sediment concentration to the main canal to 500 ppm. If the basin is maintained and operated properly, then the sediment concentration in the main canal should be less than 500 ppm. Moreover, the sediment transport capacity of the main canal was evaluated using the computer program Design of Canal for Sediment Transport (DOCSET). It was found that near the head reach (0.22 km to 9.44 km) the sediment transport capacity of the main canal was around 300 ppm. The intake of Secondary Canal S9 is at 24.79 km from the head of the main canal. Hence if the settling basin is maintained properly, the sediment inflow to Secondary Canal S9 will be 300 ppm or less.

Figures 7.15 and 7.16 show the measured sediment concentration near the head of Secondary Canal S9. The average concentration during 2004 was 148 ppm while it was significantly higher in 2005, averaging around 789 ppm. This was mainly due to the disturbances in the operation of dredgers for removing sediment from the settling basin. For comparative study, Secondary Canal S9 canal was redesigned for a sediment carrying capacity of 300 ppm (case-III) using the improved approach. However, the canal can be designed for a higher carrying capacity depending upon the requirements.

It was found that minor changes in the slope and width were sufficient to increase the sediment transport capacity to 300 ppm (Table 8.5). This canal under normal conditions would not show depositions throughout its length if constant water and sediment flow is passed.

Table 8.5   Geometry of the redesigned canal using improved approach.

Length		Discharge	Bed width	Bed slope	Water depth	Equilibrium sediment concentration
From	To					
m	m	m³/s	m	m/1,000 m	m	ppm
0	398	5.60	5.15	0.243	1.22	303
398	4,600	4.75	4.65	0.253	1.16	303
4,600	6,700	4.25	4.30	0.262	1.12	304
6,700	7,066	3.25	3.65	0.278	1.02	300

Modelling and the results

The above presented designs (three cases) were evaluated using the model SETRIC. In the first scenario, a single reach of the canal (398 m to 4,600 m) was modelled. The flow condition was assumed to be a uniform flow. A control structure was placed at the downstream boundary that would ensure a normal depth in the canal. A constant water inflow of 4.75 m³/s and a sediment inflow of 300 ppm were applied in the upstream boundary. It was assumed that the canal would be maintained ideally and there would be no change in roughness due to weed growth during the whole simulation period. Brownlie's predictor was used for computing the sediment transport capacity under equilibrium condition. The model was run for a period of 365 days. As expected, there was no deposition in the canal designed with the improved approach (case-III). The results are presented in Figure 8.10.

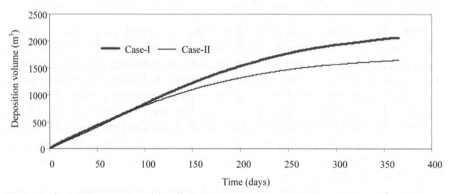

Figure 8.10   Sediment deposition with time.

As the incoming sediment was more than the carrying capacity of the canal deposition took place. Deposition started from the upstream and moved to downstream. Initially, the rate of deposition was faster and decreased with time. The canal slope slowly became steeper by sediment deposition and the carrying capacity of the section was increased. This helped to transport more sediment to the downstream. The total deposited volume in case-II after one year was around 80% of that of case-I. That means, if the discrepancy in the design could have been rectified, the deposition of sediment in Secondary Canal S9 could have been reduced by 20% under normal flow conditions.

In the second scenario, all the four reaches of the canal from 0.00 km to 7.066 km were modelled. The four reaches with different geometries and design discharges were connected by control structures. The control structures would ensure a normal water flow in the canal. The off-takes were placed at the end of each reach, so that the flow in each reach was maintained equal to the design discharge. Using a mean sediment size (d_{50}) as 0.10 mm and Brownlie's predictor, the three cases were modelled for a period of 365 days. The results are presented in Figure 8.11. The modelling results for the whole canal length (Figure 8.11) were comparable with the results of a single reach of the canal. The total deposition in case-II was around 75% of case-I. There was a little deposition in the case-III but that was negligible as compared to case-I and case-II.

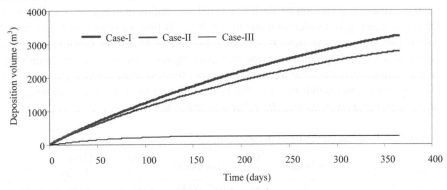

Figure 8.11   Sediment deposition with time in the canal system.

Conclusions

– canals designed based on Lacey's theory are not efficient in terms of sediment transport. It is very difficult to assess these canals, as no sediment characteristics are needed for the design;
– the existing canal is relatively too wide and as shown above, the canal sections can be narrowed. The canal designed using the improved approach is deeper as compared to the existing one. The topography is such that a deeper canal could have been easily constructed during the system rehabilitation;
– the improved approach for canal design considers the expected sediment load and sediment characteristics explicitly in the design. The roughness is determined in a rational way so that the assumed hydraulic parameters become closer to those in the actual field. This will ensure that the predicted sediment transport process is similar to the actual transport process in the field;
– minor change in bed width and/or in bed slope can change the sediment transport capacity of the canal. The canals designed by using other methods can be easily upgraded for a specific sediment transport capacity by using the improved approach;
– wider canals are not only poor in sediment transport but also are expensive from construction perspectives.

Evaluation Secondary Canal S14

Secondary Canal S14 that was designed by using an energy concept was also evaluated (ref. sub-section 5.2.2). The geometric data of the constructed canal and its sediment transport capacity under equilibrium condition is presented in Table 8.6. The other parameters of the canal are:
– side slope   = 1: 1.5
– representative sediment size (d_{50})   = 0.10 mm
– equilibrium sediment transport predictor   = Brownlie

It can be observed that the sediment transport capacity of the canal designed using the energy concept is comparable to the design of Secondary Canal S9 using Lacey's equation (Table 8.1). Besides, some inconsistency could be observed in the transport capacity along the canal. The condition for suspended sediment transport that the energy $E = \rho g V S_0$ should be constant or increasing may be a necessary but not the sufficient condition to design a canal for sediment transport (where, ρ is density of water, g is acceleration due to gravity, V is flow velocity and S_0 is friction slope). The flow velocity downstream of 3,850 m is 0.55 m/s as compared to 0.64 m/s upstream of that point, hence the transport capacity decreased even though the energy is not decreasing due to an increment in the bed slope. Hence, not only the bed slope and water depth but also the bed width and the side slope influence the sediment transport process and should be accounted for.

Table 8.6   Evaluation result of the existing design (Secondary Canal S14).

Length		Discharge	Bed width	Bed slope	Normal depth	Flow velocity	C
From	To						
m	m	m^3/s	m	m/1,000m	M	m/s	ppm
0	300	8.17	7.0	0.20	1.40	0.64	228
300	2,975	7.24	6.0	0.20	1.40	0.64	227
2,975	3,850	6.92	5.5	0.20	1.42	0.64	229
3,850	4,600	3.27	3.6	0.23	1.12	0.55	195
4,600	7,480	3.03	2.5	0.23	1.24	0.56	211

C = Equilibrium sediment concentration.

Table 8.7   Redesigned for 300 ppm transport capacity (Secondary Canal S14).

Length		Discharge	Bed width	Bed slope	Normal depth	Flow velocity	C
From	To						
m	m	m^3/s	m	m/1,000m	m/s	m/s	ppm
0	300	8.17	6.50	0.23	1.39	0.68	301
300	2,975	7.24	6.10	0.24	1.33	0.67	302
2,975	3,850	6.92	5.85	0.24	1.32	0.67	300
3,850	4,600	3.27	3.70	0.29	1.03	0.61	300
4,600	7,480	3.03	3.50	0.29	1.01	0.60	299

Table 8.8   Redesigned for 500 ppm transport capacity (Secondary Canal S14).

Length		Discharge	Bed width	Bed slope	Normal depth	Flow velocity	C
From	To						
m	m	m^3/s	m	m/1,000m	M	m/s	ppm
0	300	8.17	6.2	0.31	1.31	0.76	496
300	2,975	7.24	5.7	0.32	1.27	0.75	500
2,975	3,850	6.97	5.6	0.32	1.24	0.75	497
3,850	4,600	3.27	3.5	0.38	0.98	0.67	495
4,600	7,480	3.03	3.3	0.39	0.96	0.67	496

However, the approach used in the canal system of Secondary Canal S14 is better as compared to that of Secondary Canal S9 because in the former case the whole network has been considered as a single unit. The sediment transport capacity of Secondary Canal S14 has been related to the capacity of the main canal and similarly the capacities of the lower order canals have been related to the respective parent canals.

The same canal system was redesigned for a sediment transport capacity of 300 and 500 ppm with representative sediment size (d_{50}) = 0.10 mm and Brownlie as equilibrium sediment transport predictor. The side slope of the canal is 1: 1.5 (V: H). The results have been presented in Tables 8.7 and 8.8. It has been found that the slope of the canal has to be increased to increase the sediment transport capacity. Further, the normal water depth of the canal has also decreased due to the increase in velocity. Considering the topography of the S14 canal it would not have to be a problem to adjust the slope and water depth during the modernization. There are 13 drops in the S14 canal from 0 m to 7,480 m with different size providing a total drop in elevation of 13.05 m. The bed level of the canal at the downstream of the drop could have been easily adjusted to maintain the required water level near the head of the off-take.

8.4   Management aspects

Canals for sediment transport are designed for a specific sediment characteristic and a constant water and sediment flow under equilibrium condition. However, such conditions are seldom encountered in irrigation canals. The changing irrigation demand, different irrigation schedules to meet the demands and the flow control structures to manage those water delivery schedules are the general features of modern irrigation schemes. Thus a canal has to pass varying discharges throughout the irrigation season. Besides, if there is no proper sediment removal facility at the head, the canal has to transport various amounts of sediment. In this section, the effects of changing water demands and operation of canal with different water delivery schedules to meet those demands on sediment movement have been evaluated. The performance of the canal designed by existing methods and by the improved approach have been compared under changing flow and sediment transport conditions. For this mainly the following scenarios were considered:

–   constant irrigation water requirement with the design water delivery schedules to meet the requirements and with a constant sediment inflow rate;
–   changing irrigation water requirements and different possible modes of water delivery schedules to meet the changing demands with a constant sediment inflow rate;
–   measured irrigation supply and water delivery plans with measured sediment inflow rate;
–   changing irrigation water requirements and water delivery mode with changing but controlled sediment inflow rate;
–   changing water requirements and water delivery modes with not properly controlled sediment inflow rate.

8.4.1   Design water delivery schedule with constant water inflow

As per the design, Secondary Canal S9 has to supply irrigation water to the sub-secondary canals on a rotation basis. All the off-takes are divided into two groups and each group receives water for 3.5 days. The general layout of the off-takes and their grouping is shown in Figure 7.9. The delivery of water in rotation results in the fluctuation in discharge in the different reaches. That is, in reality, Secondary Canal S9 does not carry the design flow throughout its length. The actual flow is either more or less than the design discharge (ref. Table 8.9).

Table 8.9   Water flow along the canal as per designed water delivery plan.

Reach	Distance (m) From	To	Length (m)	Water flow (m³/s) Design	Group A	Group B	Position of off-takes and structures
	0	50	50	5.60	5.60	5.60	SS9-A, SS9-C
1	50	360	310	5.60	5.44	5.01	SS9-B
	360	398	38	5.60	4.16	5.01	WLR-1
2	398	1,114	716	4.75	4.16	5.01	Drop-1
3	1,114	1,720	606	4.75	4.16	5.01	Drop-2
4	1,720	2,242	522	4.75	4.16	5.01	Drop-3
5	2,242	3,225	983	4.75	4.16	5.01	SS9-D
	3,225	3,286	61	4.75	4.16	4.23	WLR-2
6	3,286	4,590	1,304	4.75	4.16	4.23	S9-T1, S9-T2
	4,590	4,600	10	4.75	3.50	4.23	WLR-3
7	4,600	5,910	1,310	4.25	3.50	4.23	Drop-4
8	5,910	6,620	710	4.25	3.50	4.23	SS9E
	6,620	6,700	80	4.25	3.50	3.30	WLR-4
9	6,700	7,055	355	3.25	3.50	3.30	SS9-F
	7,055	7,066	11	3.25	3.22	3.30	WLR-5

The change in flow in a section changes the B-h ratio that would change the contribution of roughness from bed and side slopes in the overall equivalent roughness. Besides, the change in B-h ratio affects the velocity distribution across the canal section and hence, the sediment transport capacity is altered. Thus, the rotation schedule introduces several changes in the canal's hydro-morphological behaviour. The effects of the rotation schedules on the sediment transport behaviour of the canal sections designed in the previous sections (case-I, case-II and case-III) have been analysed.

Modelling and results

It has already been mentioned that the existing Secondary Canal S9 (case-I) cannot deliver the design discharge to the off-takes without water level regulation even not during the design flow. The aim of modelling was to study the effect of water delivery schedules on the sediment transport behaviour of the canal. Hence, water level was not regulated and the canal was allowed to flow normally. It was assumed that no water level regulation was required to abstract the design flow by the off-takes.

The undershot type structures were placed at the end of each reach and the set-points were set such that the normal flow conditions prevailed in the reaches during the operation of the canal in the rotation schedules. The off-takes were placed at their respective locations and the gates of the off-takes were opened or closed at every 3.5 days. A constant discharge of 5.6 m³/s and a constant sediment concentration of 300 ppm were fed from the upstream boundary. The Brownlie's predictor has been used. The model was run for a period of 84 days. The modelling results are presented in Figure 8.12.

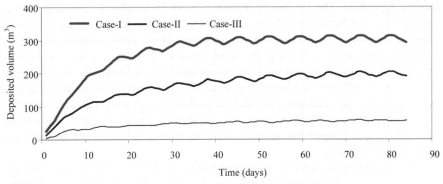

Figure 8.12   Sediment deposition with time in the whole canal system.

Figure 8.12 shows the deposition in the canal with time. The following observations and remarks can be made from the results:
-   with the water delivery schedules in place, the canal designed using the improved approach (case-III) produced less deposition than case-I and case-II;
-   the total deposition in the canal with a water delivery schedule is less than the deposition without it;
-   the deposited volume is not equal in the two water delivery modes. After some time the sediment deposited during one mode is in fact being eroded in the other mode;
-   within the modelling period, the canal reached equilibrium condition. After around 50 days there was no net increment in the deposited volume in the canal.

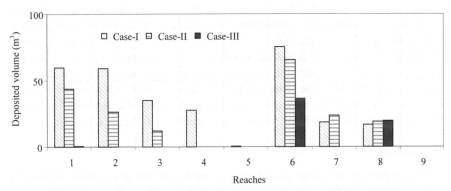

Figure 8.13   The distribution of sediment deposition in Secondary Canal S9 after 84 days.

The deposition along the canal was not evenly distributed (Figure 8.13). The deposited volume was decreasing from reach 1 to reach 5. It again increased at reach 6 and decreased in the downstream reaches. The existing canal (case-I) showed deposition almost in every reach, while the adjusted canal (case-II) showed deposition in six reaches. The canal designed with the improved approach performed well in terms of sediment transport, as the deposition was found in reaches 6 and 8.

Deposition in reach 1 was obvious, since, the carrying capacity of the canal was less than the incoming sediment concentration. The canal of both case-I and case-II showed deposition. The volume of sediment drawn by an off-take depends upon its location from the head reach if the sediment movement in the parent canal is not in equilibrium. Since, the off-takes in groups A and B are not symmetrically located and since they are not drawing the same amount of water, the total volume of sediment drawn by all the off-takes in one group keeps on changing. That is to say, the amount of sediment flowing at any section of the canal would keep on changing even though the flow for one water delivery schedule (3.5 days) remained constant. That is why the deposition amount and distribution along the canal was entirely different from that when the off-takes were drawing a fixed amount of water and sediment continuously.

The deposition in reach 6 was the highest for all the three cases. This was basically due to shortcomings in the design. This reach was designed for a discharge of 4.75 m^3/s while the flow in this reach was always less than 4.23 m^3/s. Hence, the designed section was always too large for the flow.

The canal reach between 3,286 m to 4,600 m was redesigned for a revised water discharge capacity of 4.25 m^3/s. The revised section of the canal for the three cases is presented in Table 8.10.

Table 8.10   Dimension of redesigned canal section from 3,286 m to 4,600 m.

	Bed slope	Bed width	Water depth
	m/1,000 m	m	m
Case-I	0.248	6.0	1.03
Case-II	0.248	5.2	1.03
Case-III	0.252	3.6	1.22

The canal was modelled for three cases with a changed design section for reach 6. All the input variables, assumptions and duration of modelling were taken as before. The result up to reach 6 was the same as previous, since no changes were made in that part. The result in reach 6 and beyond is presented in Figure 8.14. After the improvement the sediment transport capacity of reach 6 of the canal was increased significantly. The reduction in deposition in this reach after improvement was 100%, 41% and 84% for case-I, II and III respectively. The deposited volume in the successive reaches (7, 8 and 9) remained constant (ref. Figure 8.14 (B)).

Conclusions

Water delivery schedules introduce non-uniformity and unsteadiness in the flow of the irrigation canals. Due to the change in flow pattern, the sediment transport process of the canal is also changed. So there will be a change in the sediment transport process every time there is a change in turn or rotation of the off-takes. As shown in the modelling results, the net deposition in Secondary Canal S9 for the design discharge at each reach is more than the case when there is not always a design discharge due to the water delivery schedule. This also indicates if the effect of water delivery schedules is evaluated and if found necessary the design is modified accordingly a more efficient canal system in terms of sediment transport can be designed.

From the sediment transport perspective, it may not be a correct method to take the maximum discharge the canal has to pass as the design discharge. Such canals would be running in lesser capacity around 75% of the time and increasing the possibility of sediment deposition. Models could be used as the decision support tools in selecting the design discharge to minimize the sediment problem if the canal has to be operated in a rotation model.

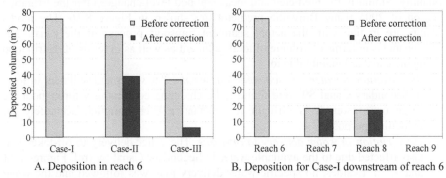

A. Deposition in reach 6   B. Deposition for Case-I downstream of reach 6

Figure 8.14 Deposition in the canal after correction.

8.4.2   Measured water inflow, sediment inflow and water delivery schedule

The performance of the proposed design (new design) with the existing water inflow, sediment concentration and system operation plan was evaluated with the existing canal and control system (existing design). The measured daily average discharge (Figure 8.8) and sediment concentration (Figure 8.12) of 2005 into Secondary Canal S9 was used. Only the sand fraction (size > 0.63 μm) of the total concentration was considered. The operation schedules and discharge through the laterals up to the water level regulator at km 7+066 is given in the appendix B. Other general parameters used for the modelling are:

- representative sediment size (d_{50})   = 0.1 mm
- sediment transport predictor   = Brownlie
- canal length   = 7,066 m
- simulation period   = 116 days

For modelling the effect of management on the canal designed by the new approach (new design) the following adjustments were made:
- the water delivery duration and discharge to the sub-secondary canals was taken similar to that for the existing design;
- the set-points upstream of the water level regulators were adjusted by linear interpolation comparing the normal water depths in existing canal and new canal for the given discharge;
- the flow control and conveyance structures were redesigned to create similar hydraulic conditions as that for the existing design (uniform flow or backwater or drawdown) for the given discharge.

Modelling and results

The modelling results are presented in Figure 8.15. For the existing design the model predicted around 3,821 m³ deposition, while the measured volume during this period was 5,620 m³. In this case the model predicted around 68% of the measured quantity. Similarly the predicted and measured bed level changes after the irrigation season of 2005 using Brownlie predictor are shown in Figure 8.16. The model results are in line with the actual sediment transport process of the canal. The discrepancies in terms of volume of total sediment as well as in terms of distribution along the canal are mainly due to:

— *the variation in water and sediment inflow rate.* The water and sediment inflows to Secondary Canal S9 were found to be changing continuously throughout the irrigation season. It is difficult to schematize such variation accurately for modelling purpose;

— *the variation in water delivery to off-takes.* The water delivery to off-takes was also affected due to the available water in Secondary Canal S9 (ref. Figures 7.11 and 7.12). Besides, no specific water delivery plan was followed and the water level regulator gates were being frequently adjusted to increase or decrease the flows to the off-takes. It is neither possible to record the water flows to off-takes nor schematize it for modelling accurately;

— *the water level regulators and drop structures.* During the rehabilitation, the existing water level regulators and the drop structures in Secondary Canal S9 were not removed but the same were used after necessary rectification and maintenance works. But the crest level of all the drops and water level regulators was increased to maintain the design set-point in the canal. The increase in crest height increased the energy of flowing water and so, the energy dissipaters provided in the downstream could not contain the jump fully within the stilling basin. This has caused erosion of the canal banks in the downstream. Heavy protection works (boulder pitching) have been provided to control the erosion, mostly without success (Figure 8.17). Hence the bed immediately downstream was eroded and the material was deposited further downstream. This created obstruction to flow and accelerated sediment deposition process in that reach. In the model it was assumed that the energy was fully dissipated and the flow was normal immediately downstream of the structure;

— *unauthorised obstruction to flow.* the farmers put illegal obstruction across the structure to raise the water level especially during night time and diverted water towards their field thus increasing the sediment deposition in the upstream.

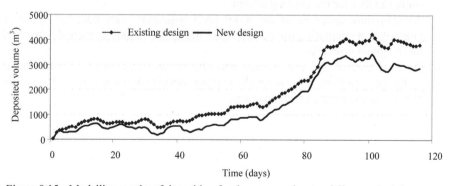

Figure 8.15   Modelling results of deposition for the measured water delivery schedules.

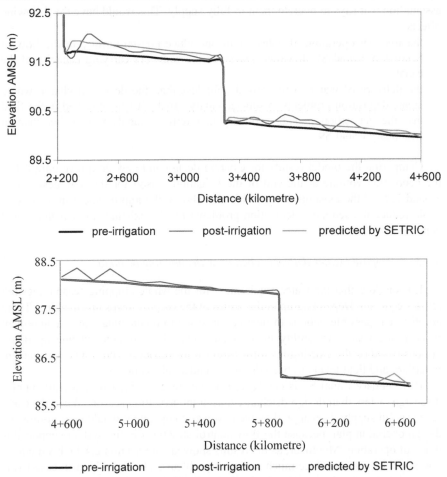

Figure 8.16 Measured and model predicted deposition along some selected stretches of Secondary Canal S9.

Figure 8.17   Protection work downstream of a water level regulator in Secondary Canal S9.

Existing water diversion to Secondary Canal S9 as well as the water delivery to sub-secondary and tertiary canals from it was not optimal. The sub-secondary and tertiary canals were operated in less than the design discharge contrary to the design assumption that says to operate them in fully supply. This would have the following effects:

– the frequent operation of gates would introduce non-uniformity in the flow of Secondary Canal S9. It would influence the sediment carrying capacity of the canal;
– the delivery of water to the off-take for less than the design discharge would create deposition problems in sub-secondary canals. It was observed in the field that the deposition in sub-secondary and tertiary canals was higher than in Secondary Canal S9.

Comparing the model results for existing design and new design (Figure 8.15), the deposition volume at the end of the irrigation season for the new design was around 75% of the existing design. This also shows, the improvement in the design could reduce the sediment deposition problems for the existing water management conditions.

8.4.3   Proposed water delivery schedule with changing demand

It is obvious that the water requirement for the whole cropping season does not remain constant. Depending upon the stages of the crops it keeps on changing. There are different possible options to operate the system on changing water requirement or changing water availability in the main canal. The possible scheme operation options based on the existing infrastructure and its impact on the water management in tertiary and field level has been discussed under sub-section 7.2.2.

If Secondary Canal S9 is considered, it will receive either design (full supply) discharge or less than design discharge during the irrigation season. When the canal receives full supply discharge, then the best way to operate it would be to follow the design operation plan because the flow control structures in the system support this mode of operation. Moreover, as the results showed (sub-section 8.4.1) this mode of operation did not introduce much sediment problems. The deposition in case-I was more related to the shortcoming in the design itself than in the water delivery mode. Further, if the canal receives full supply water on rotation basis (intermittently), the design operation mode should be preferred but the duration of off and on would have to be changed.

If, however, the canal receives less than the design discharge then there would be two possible ways of operating it. The first one would be to follow the design operation mode (rotation in two groups) and supply available water proportionally to the sub-secondary canals and the second one to rearrange the sub-secondary canals in more groups depending upon the available water and deliver full supply flow in rotation.

As observed in the field, no specific modes were being followed. The off-takes were operated randomly as per the demand of the farmers at less than the design discharge. When the demand was less then the canal could not be operated in the rotation of two groups as the available or supplied water was less then the total design capacity of the off-takes in one group.

The monthly water demand, at the head of Secondary Canal S9 canal for paddy (which is also the major crop), considering the command area of 7,921 ha is shown in Table 8.11. The demand fluctuates from 114% to about 46% of the of the design discharge.

Table 8.11   Design inflow and suggested water delivery plan for Secondary Canal S9.

Month/period	Irrigation requirement		Flow required at the head	Percentage of design flow	Rotation mode
	mm/day	l/s-ha	(l/s)	%	
Jul (period I)	4.20	0.486	3,900	69.6	3 groups
Aug (period II)	5.00	0.579	4,600	82.1	2 groups
Sep (period III)	2.70	0.313	2,500	46.4	4 groups
Oct (period IV)	7.00	0.810	6,400	114.3	2 groups

The canal is proposed to operate in the design water delivery plan up to 80% of the design discharge. Since, the flow control structures have been provided to support the design water delivery plan, it is suggested to follow the same during design flow conditions. When the flow in the canal becomes 60% to 80% of design flow, then it is proposed to divide the off-takes in three groups and provide water in rotation. Similarly, when the flow becomes 40 to 60% of the design discharge, then it is proposed to divide the off-takes in 4 groups (refer Table 8.12). It is also suggested to maintain the *on* period of one rotation group to 3.5 days, while adjusting the *off* period depending upon the number of groups in rotation. This would make it easier for the farmers to manage the tertiary and field canals as the system below sub-secondary level has been planned and developed to deliver full supply flow for 3.5 days.

The inflow sediment concentration was assumed to be constant during the whole irrigation season. The following information was used for modelling:
- mean sediment size (d_{50})   = 0.10 mm
- average sediment concentration   = 300 ppm
- equilibrium sediment transport predictor   = Brownlie
- total simulation period   = 122.5 days

The existing design (Table 8.3) as well the improved design (Table 8.4) were modelled for the above water inflow rate, water delivery schedules and sediment inflow rate. The existing control and conveyance structures as shown in Table 8.9 were placed in the respective positions. The modelling results for the both the cases are presented in Figure 8.18.

Period I and III refer to low flows. In period I the deposition was the highest. Deposition may be reduced if Secondary Canal S9 were divided into three reaches (head, middle and tail) and the off-takes in the corresponding reaches were operated one at a time. This options was not considered since, the design infrastructure did not support this mode of operation. The design flow for the canal and the control structures at any location has been approximately half the total water requirement below that point. Hence, it was not possible to arrange all the off-takes in the tail reach in one group.

Table 8.12   Proposed water delivery schedules to sub-secondary and tertiary canals of Secondary Canal S9 for the different rotation mode.

Scheduling of sub-secondary and tertiary canals

Description (Off-takes)	A	C	B	D	T1	T2	E	F	T3	T4	G	H	I	J
Design discharge in S9 (l/s)	5,600				4,750		4,250		3,250		2,800	2,300	1,800	
Position of off-takes (m)	50	50	360	3,225	4,590	4,590	6,620	7,055	9,586	9,586	11,363	12,293	14,340	14,340
Off-take design discharge	160	590	1,270	780	410	250	930	700	80	420	1,140	470	1,650	1,410
Rotation of 2 Groups														
Off-takes in Group A (l/s)	160	--	1,270	--	410	250	--	700	80	--	--	470	1,650	--
Flow in S9 (l/s)	6,400	6,240	6,240	4,970	4,970	4,560	4,310	4,310	3,610	3,530	3,530	3,530	3,060	--
Off-takes in Group B (l/s)	--	590	--	780	--	--	930	--	--	420	1,140	--	--	1,410
Flow in S9 (l/s)	6,400	6,400	5,810	5,810	5,030	5,030	5,030	4,100	4,100	4,100	3,680	2,540	2,540	--
Rotation of 3 Groups														
Off-takes in Group A (l/s)	160	590	--	--	--	250	--	700	--	--	--	470	--	1,410
Flow in S9 (l/s)	3,920	3,760	3,760	3,170	3,170	3,170	2,920	2,920	2,220	2,220	2,220	2,220	1,750	--
Off-takes in Group B (l/s)	--	--	1,270	--	410	--	--	--	80	--	--	--	1,650	--
Flow in S9 (l/s)	3,920	3,920	3,920	2,650	2,650	2,240	2,240	2,240	2,240	2,160	2,160	2,160	2,160	--
Off-takes in Group C (l/s)	--	--	--	780	--	--	930	--	--	420	1,140	--	--	--
Flow in S9 (l/s)	3,920	3,920	3,920	3,920	3,140	3,140	3,140	2,210	2,210	2,210	1,790	650	650	--
Rotation of 4 Groups														
Group A (l/s)	--	--	--	--	--	--	--	--	80	--	--	470	1,650	--
Flow in S9 (l/s)	2,800	2,800	2,800	2,800	2,800	2,800	2,800	2,800	2,800	2,720	2,720	2,720	2,250	--
Group B (l/s)	--	--	--	--	--	--	--	--	--	420	1,140	--	--	1,410
Flow in S9 (l/s)	2,800	2,800	2,800	2,800	2,800	2,800	2,800	2,800	2,800	2,800	2,380	1,240	1,240	--
Group C (l/s)	160	--	1,270	--	410	250	--	700	--	--	--	--	--	--
Flow in S9 (l/s)	2,800	2,640	2,640	1,370	1,370	960	710	710	10	10	10	10	10	--
Group D (l/s)	--	590	--	780	--	--	930	--	--	--	--	--	--	--
Flow in S9 (l/s)	2,800	2,800	2,210	2,210	1,430	1,430	1,430	500	500	500	500	500	500	--

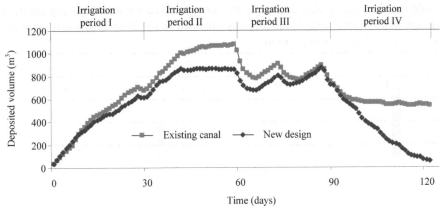

Figure 8.18   Sediment deposition along Secondary Canal S9 with time.

The canal had 80% of the design flow in period II, while the operation mode adopted was for design flow condition. Hence, deposition could be expected. However, the new design canal performed better. In period III, the total deposition was actually decreased. This clearly indicated that if planned properly, the sediment deposition could be reduced even though the canal was running at only half of the design discharge. In period IV, the flow in Secondary Canal S9 was more than the design discharge and the sediment transport capacity in almost every section was improved. Hence, the deposited material in the previous three periods was eroded in the fourth period. Comparing the results with existing design and new design, the new design performed better.

Conclusions

Depending upon the available infrastructures the available water can be diverted to the lower order canals on different modes. The water delivery modes that are easiest from management point of view may not be efficient from sediment transport perspectives. In the above analysis the proposed water delivery mode meets the irrigation requirements of different periods, is well supported by the available infrastructure and has least sedimentation problems.

The proposed water delivery schedule has been designed not only considering the sediment transport problem in Secondary Canal S9 but also in the sub-secondary and tertiary canals. The aim of the design would have to be to view the sediment transport process holistically rather than solve the problem of one point.

Figure 8.19 shows the relative sediment transport capacity of the off-takes and Secondary Canal S9 near the head of relative off-takes. It clearly indicates the shortcoming in the design and the difficulty in the management of the canal system. The off-takes not only take the water but also the sediment. It the capacity of the off-take canal is not capable of carrying the sediment load coming from the secondary canal, the deposition starts. Deposition not only reduces the water withdrawal from the secondary canal but also its conveyance capacity (Figure 8.20), which has lead to the inequitable water distribution and created dispute among the farmers. That is also the reason why the water delivery schedules were not followed. Since the canals could not draw the design discharge, the farmers asked for more opening duration. Hence, more canals were simultaneously operated in less than design discharge and again that lead to more deposition.

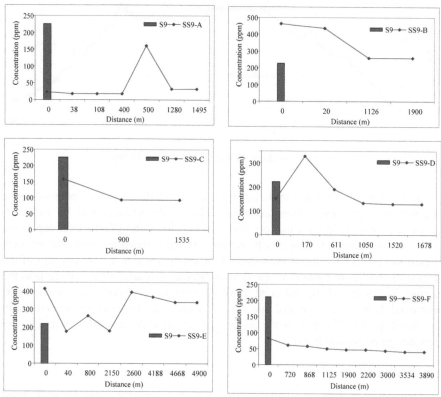

Figure 8.19   Equilibrium sediment transport capacity of off-takes and Secondary Canal S9.

Figure 8.20   Heavily deposited sub-secondary canals.

The existing poor conditions of the sub-secondary canals are due to the shortcoming in the design as well as due to the improper water delivery plans. The proposed plan ensures either full supply or no supply to the sub-secondary and tertiary canals during the whole irrigation season and this will help to reduce the sedimentation problem in sub-secondary canals. For properly addressing the problem faced by the sub-secondary canals, they need to be redesigned by using appropriate design techniques that deal with the sediment transport aspect more explicitly.

The flow velocity in major part of the sub-secondary and tertiary canal is so low that even the clay particles are settled (Figure 8.21). Since, the finer particles mostly move in suspension, they have tendency to settle in the side slopes, as the flow velocity and the shear stress is low there. The clay content in the fine sediment as well as the weed help to bind and retain the deposited material in the slopes. Hence, the sedimentation in lower order canals has two way effects; raising of bed level and narrowing of the width.

Deposition of sand and silt in the bed and clay in the slopes

Figure 8.21   Sediment deposition pattern in sub-secondary and tertiary canals.

8.4.4   *Proposed water delivery schedule with variable inflow of water and sediment*

In Secondary Canal S9, the sediment inflow rate keeps on changing. There is daily change as well as seasonal change in the sediment concentration of the river. June to October is the monsoon season, when the sediment concentration in the river becomes the highest. A settling basin has been provided at the head of the main canal to control the sediment entry to the scheme. The size of the settling basin is not sufficient to accommodate the entire sediment load coming into the system during one irrigation season from June to October. Hence, dredgers have been provided to continuously pump out the deposited volume from the settling basin. If, due to some reasons, the settling basin is not completely empty before the start of the irrigation season in June or the dredgers are not operated as per plan, the efficiency of the basin decreases and more sediment enters into the scheme. Two scenarios were used for evaluating the performance of the canal for the proposed design and water delivery schedule. Firstly the settling basin and the dredgers operating as per the design assumptions and secondly the dredgers were not operated as per assumption and the settling basin efficiency decreased with time.

Normal operation of the settling basin and the dredgers

Normally, sediment load in the river starts increasing from June and becomes the highest in August and then starts decreasing. The measured (daily average) of sediment concentration in Koshi River and in the Main Canal downstream of the settling basin for the year 1999 and 2000 is shown in Figures 8.22 and 8.23. The fluctuation of sediment load in the river directly influences the efficiency of the settling basin. Increase in sediment concentration in the river means increased loading of the settling basin. This would speed up the filling process and rate of filling becomes more than the rate of removal by dredgers. This leads to increased concentration of outflow from the settling basin.

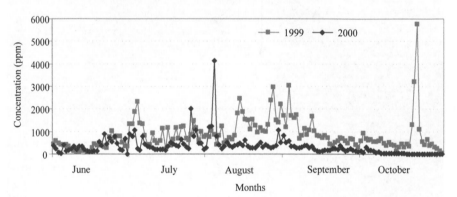

Figure 8.22   Sediment concentration (>0.063 mm) in Koshi River from June to October.

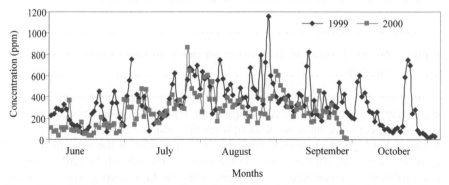

Figure 8.23   Sediment concentration (total) in Main Canal downstream of settling basin from June to October.

Under normal operating conditions of the dredgers, the sediment concentration in the canal is controlled. It normally follows the general trends in the river but the maximum concentration is expected to be less than 500 ppm. In 2004, the concentration in Secondary Canal S9 remained fairly low and became high in August following the trend of the river. While in 2005, the inflow sediment concentration was relatively high and kept on increasing with time (ref. Figures 7.16 and 7.17). It should be remembered that the operation of the settling basin and the dredgers were not optimal during 2004 and the situation worsened in 2005 due to management related problems. The expected sediment inflow rate in Secondary Canal S9 has been assumed to be as shown in Table 8.13 and the same has been used to evaluate the designs for the given water inflow and system operation plan.

Table 8.13   Expected sediment concentration (ppm) inflow to Secondary Canal S9.

Condition	July	August	September	October
Normal operation of settling basin and dredgers	300	450	400	300

Both the existing canal system (existing canal) and the canal redesigned by the improved approach (new canal) were modelled. The geometry of the canals, the positioning of the off-takes and control structures, the water delivery plan, sediment transport predictor and sediment characteristics were taken similar to that of sub-section 8.4.2. The output results are presented in Figure 8.24.

Figure 8.24 gives the modelling results of sediment transport process for the existing design and the new design in different irrigation periods. Both the cases showed a similar sediment transport trend for the water and the sediment inflow in all the irrigation periods. In period I, the deposited volume was almost the same, while at the end of the irrigation period II, the total deposition in the newly designed canal was less by around 2,000 m^3. At the end of irrigation period III, however, the net deposited volume in the canal in both the cases was same. In irrigation period IV, erosion of the deposited material started in the canal as the inflow discharge was more than the design value and the carrying capacity of the canal also increased accordingly. The increment in carrying capacity for the new design was more than for the existing design, thus at the end of the irrigation season the net volume remained in the canal was only around 16% of that for the existing design.

Moreover, the difference in final deposition in sediment volume as compared to the one with constant sediment inflow of 300 ppm (Figure 8.18) is very low. Hence, so long as the settling basin and dredgers were operated as per the design assumptions, the proposed water delivery schedule would not create a significant deposition problem.

The settling basin of SMIS, can not accommodate the sediment of one irrigation season, if the dredgers are not continuously operated. In case, the dredgers are not operated smoothly, the efficiency of the settling basin to trap sediment will decrease and more sediment will find its way into the main canal. The main canal in the initial reach has a low transport capacity and hence will get deposited first. Once the sediment crosses the upper reach of the main canal, almost all the sediment reaches the head of secondary canals, as the lower reaches of the main canal have higher sediment transport capacity.

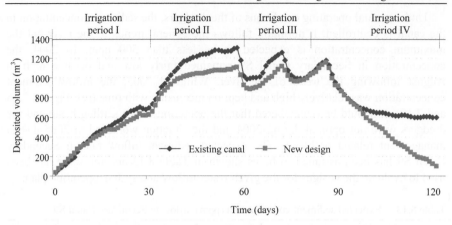

(With variable C) Dredgers working condition

Figure 8.24   Modelling result of Secondary Canal S9 for normal operation of settling basin.

Settling basin not operated as per the design

When the settling basin and the dredgers are not operated properly then the sediment concentration keeps on increasing with time. Even though the concentration in the river actually decreases towards the end of the irrigation season (October), the sediment inflow is expected to increase continuously due to decreasing efficiency of the settling basin. Considering the measured sediment concentration trends at the head of Secondary Canal S9 during 2004 and 2005, the average concentration that can be expected under this condition is as given in Table 8.14.

Table 8.14   Expected sediment concentration (ppm) inflow to Secondary Canal S9.

Condition	July	August	September	October
Dredgers not working properly	300	500	600	700

Using the similar data as used for the condition when the settling basin was functioning as per the design, the two canals were modelled and the modelling results are as shown in Figure 8.25.

The deposition when the settling basin was not operated properly (Figure 8.25) after the irrigation season was much higher (around 300%) as compared with the one with the settling basin operated as per design (Figure 8.24). Till period II, the deposition pattern was similar, since there were no significant differences in inflow sediment concentration. But in period III and IV more deposition could be observed as the incoming sediment load was increasing due to the problem in the operation of the dredgers.

In this case also, the total deposition in the newly designed canal was around 55% of the existing canal. Hence, a canal designed with the improved approach would have reduced the sediment transport problem.

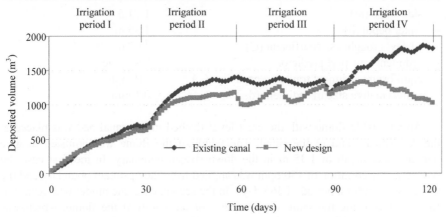

Figure 8.25   Modelling result of Secondary Canal S9 while the settling basin is not operated as per design.

Conclusion

The proposed improvement in the design and water delivery plan is tested with the fluctuation in the sediment inflow rate due to the possible problem in dredger operation. Dredgers are the key components in SMIS that largely determine the amount and quality (in terms of sediment concentration) of water to be diverted to the irrigation scheme. The model results showed that with the improved design and suggested operation plan the deposition in Secondary Canal S9 would be in the range of 1,000 m³. This is around 26% of the value predicted by the model with measured water and sediment inflow and operation mode during 2005. This also indicates, with the better understanding and application of the sediment transport concepts in the design and planning of the irrigation canals, the irrigation schemes can be operated with the reduced operation and maintenance cost.

8.4.5   Effects of flow control in sediment transport

The Sunsari Morang Irrigation Scheme has upstream flow control system in which the water level in a reach is controlled by a water level regulator (WLR) at the downstream end. The purpose of water level regulator is to maintain the target water level in the canal reach by obstructing the normal flow. Hence, the flow in the canal changes from uniform to non-uniform. In the design of canal the hydraulic and sediment transport computations are made assuming the flow to be uniform. Once the flow is non-uniform the sediment transport pattern changes. This aspect was evaluated in this section that would provide an insight on the influence of flow control in the overall result of sediment transport analysis. For the evaluation of the effect of flow control structure in the sediment transport process a canal designed with the following hydraulic and sediment characteristics was used:
– design discharge (Q)   = 7.5 m³/s
– canal length (L)   = 5,000 m
– bed width (B)   = 6.2 m
– bed slope (S$_0$)   = 0.24 m/1,000 m

- side slope (m)   = 1 : 1.5
- water depth (h)   = 1.35 m
- Chézy's roughness coefficient (C)   = 43.7 m$^{1/2}$/s
- sediment transport predictor   = Brownlie
- equilibrium sediment transport capacity (C$_e$)   = 310 ppm
- representative size of sediment (d$_{50}$)   = 0.1 mm

An adjustable flume with the crest level flushed in the canal bed was placed at the downstream end of the canal. The width of the flume was adjusted to give normal water depth of 1.35 m in the down stream boundary. In the first case the sediment concentration of 450 ppm was applied from the upstream boundary and the model was run for a period of 365 days. In the second case, the model was again run for 365 days using the same setup except for the width of the flume, which was reduced to maintain a constant water depth of 1.46 m in the downstream boundary. The depositions predicted by the model in two cases are shown in Figure 8.26.

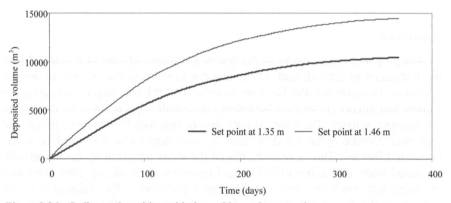

Figure 8.26   Sediment deposition with time with varying set point.

The total deposited volume in 5 km length after one year while operating the canal normally (set point at 1.35 m) was around 10,430 m^3 and the volume while operating the canal with applying a control to create a backwater condition (set point 1.46 m) was around 14,480 m^3. Thus for around 8% increment in the set-point, the increment in the deposited volume was around 39 %.

Again, the same setup was used to study the sediment transport pattern of the canal when the set-point was increased while the inflow sediment concentration remained equal to that of the carrying capacity of the canal under uniform flow condition. For this the set point was increased from 1.35 m to 1.46 m. Due to the increment in water depth the sediment transport capacity of the canal section decreased. The deposition started from the downstream end and moved in the upstream. The deposition stopped after the bed level of whole length was raised such that the backwater effect due to the increment in the set-point was changed to uniform flow. The change in bed level after 300 days has been shown in Figure 8.27.

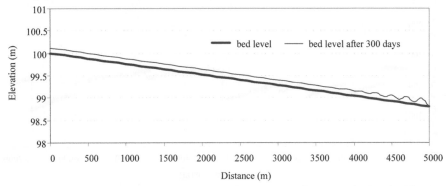

Figure 8.27   Deposition in the canal bed after 300 days when set point is raised for the same sediment concentration.

During the modernization of Secondary Canal S9 the crest level of the structures were adjusted (raised) to maintain the design water depth in the canal. The structures (flow control and conveyance) provided in the canal has been given in Table 8.9. The geometric features of these structures are given in Table 8.15.

Table 8.15   Details of the structures in Secondary Canal S9.

Location	Structure	Crest length	Crest height
m		m	m
398	WLR-1	3.00	0.103
1,114	Drop-1	3.90	0.253
1,720	Drop-2	3.35	0.166
2,242	Drop-3	3.60	0.196
3,286	WLR-2	3.00	0.122
4,600	WLR-3	3.00	0.117
5,910	Drop-4	4.40	0.285
6,700	WLR-4	2.40	0.196
7,066	WLR-5	3.00	0.136

The existing canal was evaluated with the constructed structures without any gates in the water level regulators. Only the design discharge was flowing in the canal in each reach. The sediment concentration was kept constant. The results are presented in Figures 8.28 and 8.29.

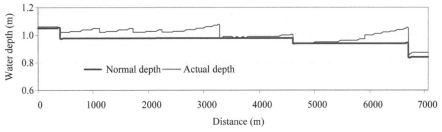

Figure 8.28   Water surface profile with the structures in Secondary Canal S9.

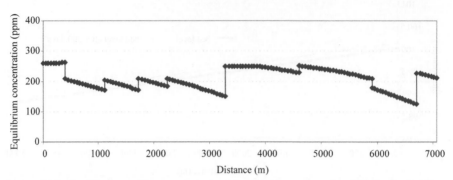

Figure 8.29   Equilibrium sediment transport capacity of Secondary Canal S9 for design flow
condition with the canal structures.

The rise in the crest levels to maintain the design water depth created backwater
in the canal during the design flow. Subsequently, the sediment carrying capacity of
the canal was reduced. In some reaches of the canal, there was no uniform flow
condition due to the back water effect. The energy dissipaters provided were not
able to dissipate the energy completely due to the rise in crest level. That was also
one of the reasons of bed and bank erosion in the downstream of stilling basin.

9
Evaluation

9.1　General

This research explored existing canal design methods in view of sediment transport processes and suggests improvements in the existing approaches considering the influence of maintenance, management (including operation) and design parameters. The aim of this research is to understand the relevant aspects of sediment transport in irrigation canals and to formulate a design and management approach for irrigation schemes in Nepal in view of sediment transport. In the process, the design methods used in the design of irrigation schemes in Nepal and their effectiveness on sediment transport have been analysed. An assessment of various design parameters that affect the hydraulic and sediment transport performance of a canal is made. The influence of non-wide canals in the velocity distribution and sediment transport is done and a correction procedure has been proposed. The impact of operation and maintenance on sediment movement has been analysed taking a case study of Sunsari Morang Irrigation Scheme in Nepal. An improved design approach for sediment transport in irrigation canals has been proposed. A mathematical model SETRIC has been improved, verified and used to study the interrelationship of sediment movement with the design and management and to evaluate the proposed design approach for irrigation canal based on the data of the irrigation scheme in Nepal.

This section presents the conclusions drawn from the research and recommendations for the future research.

9.2　Canal design aspects for sediment transport

The design of a canal is a complex process of determining its shape, slope and size based upon various aspects like amount and quality of water to be transported, type of canal to be constructed, the terrain through which it passes, socio-economic setting, climate, soil type, etc. The process becomes more complicated when the boundary of the canal is erodible and when the canal carries sediment with the water. Unwanted erosion and deposition along the canal network has become a central problem for irrigation schemes and a lot of money and efforts have been invested to find methods for a stable canal design.

Canals with a rigid boundary as well as canals carrying clear water are relatively easy to design, since in such canals the objective is to control erosion and to ensure that the velocity does not damage the lining. However, for a canal with an erodible boundary and carrying water with sediment the design needs to ensure that the velocity is high enough to convey all the sediment and at the same time it may not be so high that the bed material is eroded. Hence, a balance between the transport of sediment entering into the canal and the stability of the boundary has to be maintained. This is the most difficult aspect of canal design.

For the design of alluvial canals carrying sediment loads two approaches are in practice, namely the regime method and the rational method. The regime design methods are sets of empirical equations based on observations of canals and rivers that have achieved dynamic stability. While, rational methods are more analytical in which three equations, an alluvial resistance relation, a sediment transport equation and a width relationship, are used to determine the slope, depth and width of an alluvial canal when the water and sediment discharges and bed material size are specified.

Different empirical methods are in use at local or regional level. The fact that these methods are not being transformed to other places in the world is an indication that not all the physical parameters defining the problems are correlated by the regime methods (Raudkivi, 1990). Hence, analytical methods are better in the sense that they can be adapted for the local conditions by making suitable adjustment in the sediment transport or width predictor.

An improved rational approach is proposed for the design of alluvial canals carrying sediment load. The improvements proposed in this approach are briefly summarized in the following paragraphs.

Correction to sediment transport predictors

Most of the sediment transport predictors consider the canal with an infinite width without taking into account the effects of the side walls on the water flow and the sediment transport. The effect of the side wall on the velocity distribution in lateral direction is neglected and therefore the velocity distribution and the sediment transport are considered to be constant in any point of the cross section. Under that assumption a uniformly distributed shear stress on the bottom and an identical velocity distribution and sediment transport is considered. Irrigation canals are non-wide in nature, in majority of the cases the bed width to water depth ratio being less than 8 (Dahmen, 1994). Hence the effect of side walls on the shear stress distribution would be significant and hence in the velocity distribution across the canal. Hence, the assumption of uniform velocity and sediment transport capacity across the section and expressing them per unit width of canal does not hold true for irrigation canals. The majority of irrigation canals are trapezoidal in shape with the exception of small and lined canals that may be rectangular. In a trapezoidal section the water depth changes from point to point in the section and hence the shear stress. The effect would be more pronounced if the bed width to water depth ratio is low (which is the case of irrigation canals). In the above presented background a correction factor to the sediment transport predictor has been suggested that would improve the predictability of the predictor for the flow conditions in an irrigation canal. It has been shown by available data in this research that the correction factor increased the predictability of the predictors.

Roughness prediction and equivalent roughness

In alluvial canals there exist two stages of flows, one when there is no movement of bed material and the other when bed material is moving. No movement of bed material can be compared with the condition of a rigid boundary canal having an equivalent roughness related to the representative bed material size (d_{50}). The resistance to flow in a movable bed consisting of sediment is mainly due to grain roughness and form roughness. Grain roughness is generated by a skin friction force and form roughness by a pressure force acting on the bed forms. Since the bed forms

change continuously with the flow parameters (velocity, depth), the bed roughness also changes. There are two approaches to estimate the bed roughness:
- methods based on hydraulic parameters such as mean depth, mean velocity and bed material size;
- methods based on bed form and grain-related parameters such as bed form length, height, steepness and bed-material size.

The method based on bed form and grain related parameters is proposed to use for computing the roughness in the bed.

Earthen canals carrying sediment load will have different roughness in the bed and side. The side roughness is determined by the condition of the canal, material, etc. and will keep on changing with time depending upon the type and rate of possible weed growth in the bank, the maintenance plan and the erosion and/or protection work during canal operation. The bed roughness, however, is determined by the size of the bed material and the bed forms. The bed forms keeps on changing with the flow condition in the canal. Hence, the bed roughness is dependent on the changing flow condition during the irrigation season.

Determination and use of roughness is critical for the design of irrigation canals since it would not only affect the predicted hydraulic parameters but also would have influence on the predictability of the sediment transport equations. Mendez (1998) proposed to compute the equivalent roughness of section and applied a correction for the water depth. The method was evaluated and compared with other existing methods of computing sediment transport from the available data. It has been found that the method proposed by Mendez (1998) gave the best result for the conditions that are found in irrigation canals.

B-h ratio

Selection of a B-h ratio is one of the aspects that influence the design of irrigation canals. From flow control aspects too deep canals are not preferred in irrigation because a change in flow affects more in the depth. However, from sediment transport aspect deep canals have a higher suspended sediment transport capacity due to the higher average velocity. Further, bed load transport capacity of a deep canal is better as compared to a wide canal, because the bed load transport capacity increases linearly with bed width but exponentially with the depth (Dahmen, 1994). Based on the bed width to water depth ratio in the major irrigation canals in the world carrying clear and sediment laden water, a B-h ratio selection criteria has been proposed.

A canal design program DOCSET (Design Of Canal for Sediment Transport) has been prepared for the improved approach including the above mentioned improvements. The program can be used to evaluate the existing design for given water flow and sediment characteristics.

9.3   Mathematical modelling aspects of canal design

It is not possible to design a canal that will be non-silting and non-scouring for all discharges and sediment concentrations. Best possible solution is to balance the total erosion and deposition in one crop calendar year. Hence deposition/erosion should be allowed during the different irrigation periods. Hence, a design concentration may not be the maximum sediment concentration expected during the irrigation season. A balance has to be found between the variation in irrigation requirement and the corresponding variation in sediment concentration in the flow together with the water delivery schedules and the flow control. The best way to evaluate a canal under such scenario is to use a suitable sediment transport model.

Roughness keeps on changing throughout the canal operation. However, the canals have to be designed by taking the average roughness expected during the irrigation season. Modelling provides an option for a precise representation of these changes during the irrigation season that will increase the reliability of predicted morphological change and help in a better design from sediment transport perspective.

A flow control system is needed in an irrigation scheme to manage the water flows at bifurcations to meet the service criteria and standards regarding flexibility, reliability, equity and adequacy of delivery. A flow is regulated through water level control, discharge control, and/or volume control that make the flow non-uniform. For flows other than the design values, the gates are operated to maintain the set-point and diverting desired water to the laterals. This creates a drawdown or backwater effect and non-equilibrium sediment transport conditions. The canals are designed assuming a steady and uniform flow and an equilibrium sediment transport condition. The sediment transport equations used in the design are not capable of predicting the sediment transport behaviour under non-equilibrium conditions. Sediment transport models provide an option for predicting the sediment transport process in time under changing flow conditions. Hence, a design would have to be evaluated by using a sediment transport model and necessary changes be made to reduce the erosion/deposition.

9.4   The Model SETRIC

The sediment transport model SETRIC (Mendez, 1998, Paudel, 2002) is a one-dimensional model where water flow is schematized as quasi-steady and solved as gradually varied flow. The sub-critical flow profile is solved using the predictor-corrector approach. Roughness in the bed and sides of the canal is calculated separately. The model calculates the roughness in the bed using the Van Rijn (1984C) method, which is based on flow conditions and on the bed form and grain related parameters as bed form length, height and sediment size. Then the equivalent roughness is computed taking into account the side-wall effect as proposed by Mendez (1998).

The sediment continuity equation is solved numerically by using the Modified Lax scheme. A depth integrated convection-diffusion model (Galappatti, 1983) is used to predict the actual sediment concentration at any point under non-equilibrium conditions. For the prediction of the total sediment transport under equilibrium condition an option to select one of the three predictors, namely Ackers and White (HR Wallingford, 1990), Brownlie (1981) and Engelund and Hansen (1967) has been provided. The predicted sediment transport capacity is corrected for the B-h ratio and side slope for non-wide irrigation canals.

The theoretical background and assumptions made in the model has been discussed and the model has been used for analysis purpose. The findings on the use of model can be summarized as follows:

- *model verification.* The hydraulic and sediment transport predictions of the model have been found to be comparable with other models that are being used for research and design fields. Moreover, the model predictions are in line with the natural process of morphological changes due to the change in water flows and sediment loads;

- *model validation.* The model has been validated using measured field data as input variables and comparing the predicted morphological change results with the measured values. The model predictions have been found to be in line with the observed deposition pattern in the canal;

- *application to the field data.* The model has been used to simulate the sediment transport process in Secondary Canal S9 of SMIS using measured water flow, sediment concentration and water delivery schedules of the irrigation season 2004 and 2005 as input variables. In terms of deposited volume the model predicted 52% and 68% of the measured quantity in 2004 and 2005 respectively;

- *general use of the model.* This model can be used as a management tool. With the model it is possible to predict the sediment transport behaviour of the canal for particular water requirements and delivery schedules. This not only helps to decide how much sediment would be deposited or eroded, but also to determine what would have to be the capacity or efficiency of the sediment removal facilities and how they would have to be operated. Hence even for a canal system that has been designed without considering any sediment transport criterion, this model can help to select a better operation plan that gives minimum deposition in the system. However, the most suitable operation plan may not match with the original design assumptions.

9.5   Present canal design methods for sediment control in Nepal

The problem of sediment in irrigation schemes in Nepal is well recognized and accordingly different methods of canal design to deal with the sediment deposition and erosion have been recommended in the design manuals prepared by the Irrigation Department and are in practice. The manuals recommend:

- to use the tractive force method for unlined earth canals carrying relatively clear water;

- to satisfy both non-silting and non-scouring criteria for unlined canals carrying a sediment load. They recommend to use Lacey's regime equations, White-Bettess-Paris tables or similar equations/charts for controlling sediment deposition and tractive force equations for preventing erosion;

- to use Lacey's equations only to the areas where the sediment size and concentration is expected to be similar to those implicit in the formulae;

- to carry out a more thorough assessment of the sediment balance using quantitative formulae for large canals. Use of Engelund and Hansen or Ackers and White equation is recommended for computing the sediment transport capacity.

No specific standards have been developed so far, hence the constants and coefficients in empirical equations are selected on personal judgement. The side slope computed from Lacey's equations is steeper than normally required from slope stability considerations.

Use of Lacey's equations for canal design is difficult to justify as the sediment size and concentration generally found in the irrigation canals of Nepal are not comparable to that from which the equations have been derived.

The White-Bettess-Paris tables are derived from alluvial friction equations of White, Paris and Bettess (1980) and the sediment transport equations of Ackers and White (1973). No records regarding the use of this method for the design of canals was found and hence its performance in terms of sediment transport could not be verified. However, from evaluation with the data of SMIS, it has been found that the Ackers and White sediment transport equations over-predict the sediment transport capacity of a canal. That means that the White-Bettess-Paris tables would result in a canal with a flatter slope than actually required to carry the type of sediment prevailing in SMIS and other similar irrigation schemes of Nepal.

It has been observed that the sediment load found in the Terai canals is mostly fine ($d_{50} < 0.20$ mm). For this sediment size, the Ackers and White equations over-predict the sediment transport capacity. The Brownlie and Engelund and Hansen equations are more suitable for the type of sediment that is being found in SMIS and other irrigation schemes in Nepal.

9.6   Modernization aspects of Sunsari Morang Irrigation Scheme

Canal design for sediment control

The SMIS has been in operation since 1968 and has enough information on the type and amount of sediment being deposited yearly in the canal network. However, it has been found that the available information was not fully utilized for the redesign of canals during the modernization stages.

In practice, the design of earthen canals for sediment transport has been mostly limited to the use of Lacey's equations. The older canals have been designed using Lacey's equations while the newer ones use Lacey's formulae to compute the bed slope and use the tractive force method to check the slope for erosion control. In one of the secondary canals of SMIS, non-decreasing energy with the tractive force concept to control erosion has been used. Field observations show that these approaches of design are not able to deal with the sedimentation problems.

In Secondary Canal S9, the sub-secondary and lower order canals have been designed using a simple resistance equation (Manning's equation). It clearly indicates that the effort of designing a non-silting and non-scouring canal until now is limited to the main and secondary canals.

Irrigation management and sediment transport

The water delivery plans have been prepared entirely on the basis of water requirements and no consideration on sediment movement was made. However, the evaluation of the water delivery plans showed that the design water delivery plan did not increase the deposition problem. No specific plans existed for operating the canals for less than the design discharge. It has been found that the system is being operated more or less randomly, with no specific irrigation schedule. Overall water balance in the design shows that SMIS can not meet the irrigation demand of the whole 58,000 ha command area. Still, in both irrigation seasons of 2004 and 2005, Secondary Canal S9 received more than the design volume of water for one irrigation season (around 125 and 160% respectively). This was possible partially because, not all the target command area of 58,000 ha was being supplied with the fair share of water due to the lack of proper water distribution facilities in some part of the command area.

From sediment transport perspective, the canal would have to be operated such that during most of the time there is the design discharge in a canal and least obstruction is applied to the flow. Obstruction to flow is unavoidable in case of SMIS due to the water diversion and distribution arrangements. In such condition, the effort needs to be to design a water delivery schedule that can be implemented with a least control. Considering the water management practices being followed in the Secondary Canal S9, no design methods can prevent the deposition in the canal network.

The existing poor conditions of the sub-secondary canals are due to not only the shortcomings in the design, but also the improper water delivery practices. The design proposes either full supply or no supply to the sub-secondary and tertiary canals during the whole irrigation season; however, no specific plans for the same have been given. In reality more canals are being run for less than the design capacity. Thus causing more deposition.

Planning the irrigation canal system for sediment transport

It has been observed that the design of Secondary Canal S9 and its system lacked a proper planning from sediment transport point of view during modernization. An evaluation of the relative sediment transport capacity of the sub-secondary canals and Secondary Canal S9 near the head of the corresponding sub-secondary canals clearly indicates the shortcoming in the design and the difficulty in the management of the canal system. The off-takes for the sub-secondary canals not only draw the water but also the sediment. If the capacity of the sub-secondary canal is not capable of carrying the sediment load coming from the secondary canal, the deposition starts.

The sub-secondary canals off-taking from Secondary Canal S9 have different sediment transport capacities. Even within the same canal the transport capacity is different in the different reaches. It is not always possible to design a canal network with a constant sediment carrying capacity. However, a clear holistic approach is needed while planning the development of a canal network. The major problems due to the present planning are:

- one canal is heavily deposited as compared to others leading to unfair distribution of water. Moreover, the farmers under one canal have to invest more for maintenance than the others. Normally these canals are not cleared completely and the farmers prefer to request for longer opening duration or use other means to acquire water rather than invest more for the sediment removal;

– one reach has more deposition than the other reaches leading to disputes among
  the farmers from different reaches due to the differences in interests;
– difficult to follow water delivery plans since, the canals are not properly
  maintained and can not draw the design discharge.

A sediment deposition pocket near the head of a sub-secondary canal is one
alternative that can be adopted to avoid such problems. This would reduce the
uneven distribution of sediment load along the canal network and moreover, the cost
of removing and managing the sediment would be less. From management
perspective also this would reduce the conflicts as clearance of the sediment pocket
becomes the common responsibility of all the farmers.

9.7   Application of the improved approach for the design in SMIS

Evaluation of design for uniform conditions

The proposed improvement in the design approach of canals for sediment
transport has been evaluated and compared with the design methods used in the
canal systems of SMIS.

Secondary Canal S9 has been designed using Lacey's equations. The canal was
evaluated for its sediment transport capacity in uniform flow conditions using
Brownlie's sediment transport predictor, which is around 230 ppm. This transport
capacity is less than the sediment load expected in Secondary Canal S9. It was found
that the sediment transport capacity of the canal can be improved by around 32% by
adjusting the bed width and bed slope. The changes in bed width and slope are not
significant and could have been easily implemented during modernization of the
system.

Another Secondary Canal S14 of SMIS has been designed using the energy
concept to control the sediment deposition and tractive force concept for the control
of erosion. Evaluation of the canal revealed that the sediment transport capacity of
this canal is around 230 ppm near the head reach and decreases towards the tail. In
this case also the transport capacity of the canal is less than the expected sediment
load. Besides, the transport capacity is neither constant nor increasing in the
downstream. Hence, the criteria used to test the non-silting conditions have been
found to be insufficient.

However, the approach used in the canal system of Secondary Canal S14 is
better than that in the canal system of Secondary Canal S9 in the sense that here, the
whole system has been considered as one unit. The sediment transport capacity of
the Secondary Canal S14 has been related to the capacity of the Main Canal and
similarly the capacities of the lower order canals have been related to the respective
parent canals.

The major constraint of both the methods is that both do not use sediment size
and concentration/load explicitly in the design. The conclusions of the analysis are:
– methods that use sediment characteristics implicitly in the design may be
  sufficient for certain sediment size and load. These methods may be helpful
  when there is limited information on the type and amount of sediment load to be
  transported by the canal system;

- the sediment transport capacity is not only a function of the bed slope and water depth as assumed in the energy concept but also of the bed width and the side slope of the canal;
- sediment transport is a complex process and the method that uses the variables (both hydraulic and sediment) more explicitly will be easier to test and improve or modify to make it suitable for use in local conditions.

System management for sediment transport

The model SETRIC was used to study the effect on sediment transport process due to the system management activities namely, the change in water demand and supply, the water delivery modes based on the available water and the change in sediment load due to the variation in sediment inflow from the river or problems in proper operation of the settling basin. The improved canal design approach was evaluated comparing the results with the existing design of Secondary Canal S9.

Constant water and sediment inflow with design water delivery schedule

The model showed limited deposition in the canal for a constant water inflow of 5.6 m^3/s and a sediment inflow of 300 ppm with the design water delivery schedule. The improved design reduced the net deposition in the system by around 81% as compared to the existing design. The model results also showed that after certain time all the sediment entering into was flowing out of the system. The following conclusions can be drawn from the observation of the modelling results:
- the water delivery schedule changes the flow pattern and also the sediment transport process;
- the change in sediment transport process does not necessarily always increase sediment deposition in the canal. Sediment deposited in one rotation turn can be eroded in the other thus creating stability within the rotation schedule;
- water delivery schedules can be designed and implemented to reduce the erosion/deposition problems of a certain reach even after the system has been constructed and put into operation.

Measured water inflow, sediment inflow and water delivery schedule

The measured data of water and sediment inflow as well as the operation of Secondary Canal S9 in 2004 and 2005 for paddy were used to model the sediment transport prediction. The model predictions of morphological change in terms of volume deposition are around 52% and 68% in 2004 and 2005 respectively. The following conclusions are drawn from the study of field data and the model results:
- the sediment inflow rate in 2005 was higher than in 2004 and it was due to the problem in the operation and maintenance of dredgers used in the settling basin;
- if the settling basin is not maintained properly the sediment inflow into Secondary Canal S9 will be much higher than its transport capacity and more deposition can be expected;
- the design operation plans and assumptions have not been followed in Secondary Canal S9. From sediment transport perspective, the existing water management practices resulted in more deposition to the sub-secondary and tertiary canals than the secondary canal.

Variable water but constant sediment inflow with proposed water delivery plan

The water delivery to an irrigation scheme can not be constant throughout the irrigation season. The model was used to study the effect on sediment transport due to change in water supply by preparing a water delivery schedule for the design water requirement of the scheme during the irrigation season. Assuming a constant sediment inflow rate of 300 ppm, the sediment deposition was found to be lower than the existing deposition in the canal. The model results in terms of volume for improved design was only 10% of the results for the existing design. The following conclusions can be drawn from the modelling results:

- the periodic change in the demand and the corresponding change in sediment transport capacity of the canal can be optimized to arrive at the seasonal balance in the sediment deposition. in one period there may be deposition but that can be eroded in the next period;
- the proposed water delivery plan covers discharge fluctuation in Secondary Canal S9 from around 46% to 114%, hence during most of the time the proposed water delivery schedule can be followed;
- the proposed water delivery plan is based on the existing canal design and the available infrastructure in the system. Hence, it can be implemented in the system;
- the proposed delivery schedule ensures either full supply or no supply to the sub-secondary canals which have been designed based on the same principle. This would reduce the existing deposition problem faced by these canals.

Variable water and constant sediment inflow with proposed water delivery plan

Field measurements have shown that the sediment concentration keeps on changing during the irrigation season. The same setup with the changing water inflow and the proposed water delivery plan was used in the model to study the effect of changing inflow concentration on the overall performance of the system in terms of sediment transport. In the first case it was assumed that the settling basin would function as per design assumption and the sediment inflow during the irrigation season would be controlled. Assuming a varying average monthly sediment inflow based on sediment concentration trends in the river and in the main canal, the modelling results revealed that if the proposed operation plans were followed, then the net deposition in the canal system after the irrigation season would not increase by more than 10%.

If, however, the settling basin could not be operated properly more sediment would be flowing into the system. For a possible monthly average sediment inflow, the deposition in Secondary Canal S9 may increase up to 300%. The canal designed with the improved approach could still reduce the deposition by around 45%. The following conclusions can be drawn from the results:

- due to the continuous increase in the sediment concentration the deposited volume could not be eroded in any periods to reduce the volume;
- proper operation of the settling basin is crucial for the sustainability of the SMIS;
- the efficiency of Secondary Canal S9 to transport sediment when the sub-secondary canals are rotated in three groups is the lowest. Arranging the off-takes from head middle and tail in three groups and rotating them could improve the sediment transport capacity. However the present design and the infrastructure do not allow for such arrangement;

– the secondary canals need to be operated in rotation when there is less demand or less available water in the main canal. This would ensure design flow in the secondary canals and reduce the sedimentation problem. The main canal would have to be analysed for best mode of rotation from sediment transport and water delivery perspective.

It is possible to reduce the sediment deposition problem by proper design and management of the system. Water management may be an effective way of controlling sediment. The sediment deposition to some extent can be minimised by the canal operation plan. The sediment entering into the system has to be deposited somewhere in the system. It may be the main canal, secondary canal, tertiary canal or farm fields. Even if the system is not designed efficiently for sediment transport, it is possible to reduce an erosion/deposition problem by following certain operation mode. For example, always supplying full supply to the sub-secondary and tertiary canals of Secondary Canal S9 will reduce deposition problems there, but will increase the problem in Secondary Canal S9. Now the choice has to be made by the management where they want to concentrate the deposition.

9.8   Effects of flow control in sediment transport

One of the main features that make irrigation canals different from rivers in terms of hydraulics is the presence of flow control structures. These structures obstruct the flow to regulate the water levels. Modern irrigation schemes are more and more demand driven. That means frequent change in the flow and level is needed to meet the demand. That makes the flow non-uniform and accordingly sediment transport property is changed. The model SETRIC was used to evaluate the effect of flow control on the sediment transport process and also the status of the present flow control structures provided in Secondary Canal S9. The following conclusions can be drawn from the modelling results:
– a backwater profile created due to the increment in the set point by 8% can increase the sediment deposition by 39%. The percentage, however, depends upon the width and length of canal. Nevertheless, it shows the influence of flow control in the sediment transport process;
– almost all the water level regulators and drop structures provided in Secondary Canal S9 created backwater effect during design flow condition.

9.9   General observations and recommendation for further research

The sediment transport problems in irrigation schemes are not purely a design issue and hence, a designer should have a proper vision and knowledge of the operation and management limitations. For example, the vortex tube sediment excluders provided in SMIS had to be closed completely, because the operational limitations of the scheme were not fully analysed while installing the excluders. Similarly, the manager should have the understanding of the design concept that has been used for the scheme. No design can eliminate the sediment transport problem if the system is not operated as per the design assumptions.

The objective of a canal design should be clear, since designing the canal network that conveys all the sediment down to the field, may keep the canal free of sediment deposition but at the same time the sediment will be accumulated on the agricultural fields every year that might affect adversely the soil quality and productivity.

The sediment transport problems of an irrigation scheme, in many cases, can be avoided or reduced by modifications in the operation and management plans. Mathematical models are the helpful tools that can be used by the water management engineers to look for the necessary improvements/changes in their management plans. Hence, improvements in the model SETRIC is recommended to make it more user-friendly so that the designers and managers can easily use it and the decision makers understand it.

In most cases, the off-takes are located near the water level regulators. The flow condition near these regulators is seldom uniform due to the operation of the gates to maintain the target set-points. Accordingly the sediment transport process near the off-takes is also mostly in non-equilibrium conditions. The flow tries to adapt to equilibrium condition by entrainment or deposition of sediment, so the concentration profile keeps on changing with time. Hence, there is less chance to have a defined suspended sediment distribution ratio between the off-takes and the parent canal.

While modelling the irrigation canals for sediment movement the aspects of the change in the flow conditions and the effects of this change in the sediment transport process is important. The change in the flow conditions may be either due to the operation of the gates to maintain the water delivery schedules or due to changes in water demand and/or supply. The unsteadiness in the flow due to the operation of gates is mostly of short duration as compared to the duration of the irrigation season and the effect of this unsteadiness on the morphological change from seasonal perspective is insignificant. Moreover, the unsteadiness due to the variation in demand and supply are mostly gradual and such variations can be schematized as quasi-steady condition for sediment movement analysis. Hence, the schematization of the flow in irrigation canals as quasi-steady for modelling purpose is valid and sufficient.

Design or evaluation of an irrigation scheme for sediment transport becomes difficult due to the unavailability of data. The field measurement procedures are cumbersome and the desired level of accuracy is difficult to achieve. Hence, the designers prefer to avoid the analysis of sediment transport process. At the most they use empirical equations that require least information on sediment. Out of more than 10 large scale irrigation schemes (command area > 8,000 ha) in Nepal; very few have equipments and laboratories for sediment sampling and analysis. Among them only SMIS has established practices of taking sediment samples regularly during the canal operation.

The coarser fraction of the sediment is mostly controlled at the headwork and settling basin of an irrigation scheme. The sediment that is encountered in main and secondary canal is generally fine sand. Most of the silt fraction (sediment < 63 μm) is transported to lower order canals and the fields, where it gets deposited. The transport and deposition process of fine cohesive sediment is different from the non-cohesive fraction and its mathematical representation for modelling purposes needs additional investigations. A research in the field of cohesive sediment mainly focusing on the following topics will be helpful to extend the research on sediment transport in irrigation canals for solving practical problems:

- determination of threshold shear stress for the deposited fine sediment;
- investigation of the effect on the sediment movement process due to the turbulence created by the flow control structures that are generally provided at short intervals in the lower order canals;
- investigation of the deposition and erosion process on the bed and side slopes.

References

Abbot, M. B. and Cunge, J. A. (Eds.), 1982. Engineering applications of computational hydraulics, Volume I. Pitman Books Limited, Massachusetts.

Ackers, P., 1993. Sediment transport in open channels: Ackers and White update. Technical note No. 619, Proc. Instn. Civ. Engrs. Wat. , Marit. & Energy, United Kingdom.

Ackers, P. and White, W. R., 1973. Sediment transport: new approach and analysis. Journal of Hydraulic Engineering, ASCE, 99(11).

Agrawal, G. R., 1980. Resource Mobilization in Nepal. Centre for Economic Development and Administration, Tribhuwan University, Nepal. p. 251.

Ankum, P., 1993. Canal storage and flow control methods in irrigation. Proc. 15th Congress on Irrigation and Drainage. The Hague. 1-B, question 44, 663-679.

Ankum, P., 2004. Lectures notes on Flow Control in Irrigation Systems. IHE. Delft.

Armanini, A. and Di Silvio, G., 1988. A one dimensional model for the transport of a sediment mixture in non-equilibrium condition. Journal of Hydraulic Research, 26(3): 275-292.

Bagnold, R., 1966. An approach to the sediment transport problem from general physics. Geological Survey Prof. Paper 422.I, Washington, USA.

Bakker, B., Vermaas, H. and Choudri, A. M., 1986. Regime theories updated or outdated. Paper presented at the 37th International Executive Committee Meeting of the International Commission on Irrigation and Drainage. Lahore, Pakistan.

Belaud, G. and Baume, J.-P., 2002. Maintaining Equity in Surface Irrigation Network Affected by Silt Deposition. Journal of Irrigation and Drainage Engineering, ASCE, 128(5).

Blench, T., 1957. Regime behaviour of canals and rivers. Butterworths Scientific Publications. London, England.

Bos, M. G. (Ed.), 1989. Discharge measurement structures (Third ed.). International Institute for Land Reclamation and Improvement. Wageningen, the Netherlands.

Brebner, A. and Wilson, K. C., 1967. Determination of the Regime Equation from Relationship for Pressurised Flow by Use of the Principle of Minimum Energy Degradation. Proc. Institutions of Civil Engineers. 36, 47-62.

Brownlie, W. R., 1981. Prediction of flow depth and sediment discharges in open channels. Report No. KH-R-43A, W.M. Keck Laboratory of Hydraulic and Water Resources, California Institute of Technology, California.

Bureau of Reclamation, 1987. Design of small dams. U.S. Department of the Interior Bureau of Reclamation. US Government Printing Office, Washington, D.C. 20402.

Bureau of Reclamation, 2001. Water Measurement Manual: A Guide to Effective Water Measurement Practices for Better Water Management. U.S. Department of the Interior Bureau of Reclamation, U.S. Government Printing Office, Washington, DC 20402.

Carsten, M. R., 1966. An Analytic and Experimental Study of Bed Ripples Under Wake Waves. Georgia Institute of Technology, School of Engineering, Atlanta.

Celik, I. and Rodi, W., 1984. A Deposition-Entrainment Model for Suspended Sediment Transport. Report No. SFB 210/T/6, University of Karlsruhe, West-Germany.

Celik, I. and Rodi, W., 1988. Modelling suspended sediment transport under non-equilibrium situations. Journal of Hydraulic Engineering, ASCE, 114(10): 1157-1191.

Central Bureau of Statistics, 2006. Agricultural Census Nepal, 2001/02. Monograph, National Planning Commission Secretariat, Government of Nepal.

Chang, H., 1988. Fluival processes in river engineering. John Wiley & Sons. New York, USA.

Chang, H. H., 1980. Stable Alluvial Canal Design. Journal of Hydraulic Division, ASCE, 106(5): 873-891.

Chien, N. and Wan, Z., 1999. Mechanics of Sediment Transport (J. S. McNown, Trans.). ASCE Press. USA.

Chitale, S. V., 1966. Design of Alluvial Channels. Proc. 6th Congress of ICID. Delhi. Q20, R.17.

Chitale, S. V., 1996. Coordination of Empirical and Rational Alluvial Canal Formulas. Journal of Hydraulic Engineering, ASCE, 122(6).

Chiu, C.-L., Jin, W. and Chen, Y.-C., 2000. Mathematical Models of Distribution of Sediment Concentration. Journal of Hydraulic Engineering, ASCE, 126(1).

Chow, V. T., 1983. Open channel hydraulics. Mc Graw Hill International Book Company. Tokyo, Japan.

Clemmens, A. J., Bautista, E., Wahlin, B. T., et al., 2005. Simulation of Automatic Canal Control Systems. Journal of Irrigation and Drainage Engineering, ASCE, 131(4).

Clemmens, A. J., Holly, F. M. and Schuurmans, W., 1993. Description and evaluation of Duflow. Journal of Hydraulic Division, ASCE, 119(4): 724-734.

Clemmens, A. J., Wahl, T. L., Bos, M. G., et al., 2001. Water Measurement with Flumes and Weirs. ILRI Publication No. 58, International Institute for Land Reclamation and Improvement, Wageningen, the Netherlands.

Colby, B. R., 1957. Relationship of unmeasured Sediment Discharge to Mean Velocity. Trans. Amer. Geophy. Union, 38(5).

Constandse, A. K., 1988. Planning and creation of an environment. IJsselmeer Polders Development Authority, Lelystad, the Netherlands.

Cunge, J. A., Holly, F. M. and Verwey, A., 1980. Practical aspects of computational river hydraulics. Pitman Publishing Limited. London, UK.

Dahmen, E. R., 1994. Lecture Notes on Canal Design. IHE, International Institute for Hydraulic, Infrastructure and Environment Engineering. Delft, the Netherlands.

Dahmen, E. R., 1999. Irrigation scheduling. IHE, International Institute for Hydraulic, Infrastructure and Environment Engineering. Delft, the Netherlands.

Danish Hydraulic Institute, 1993. Mike 11: Technical reference guide. Danish Hydraulic Institute, Copenhagen, Denmark.

Danish Hydraulic Institute, 2002. MIKE 11- A modeling system for rivers and channels. Software Reference Manual,, Danish Hydraulic Institute (DHI), Copenhagen, Denmark.

Danish Hydraulic Institute, 2009. MIKE 21C River Morphology, A Short Description. Retrieved October 14, 2009 from:
http://www.dhigroup.com/~/media/E5DE3B243EFA495F9B3B510E943C67BE.ashx

De Vries, M., 1975. A morphological time scale for rivers. Paper presented at the XVIth IAHR Congress. Sao Paulo.

De Vries, M., 1985. A Sensitivity Analysis Applied to Morphological Computations. No. 85-2, Delft University of Technology, Dept. of Civil Engineering, Delft, the Netherlands.

De Vries, M., 1987. Morphological Computations: Lecture Notes F10A (1987-version). Delft University of Technology, Faculty of Civil Engineering. Delft, the Netherlands.

De Vries, M., 1993. River Engineering. Lecture notes, Faculty of Civil Engineering, Delft University of Technology, Delft, the Netherlands.

De Vries, M., Klaassen, G. J. and Struiksma, N., 1989. On the use of movable bed models for river problems: state of the art. Proc. Symposium on River Sedimentation. Beijing, China.

Delft Hydraulics, 2007. Delft Hydraulics Software: Delft3D. Retrieved May 06, 2007 from:
http://delftsoftware.wldelft.nl/index.php?option=com_content&task=blogcategory&id=13&Itemid=34

Delft Hydraulics and Ministry of Transport Public Works and Water Management, 1994a. SOBEK, technical reference guide. Delft Hydraulics, Delft, the Netherlands.

Delft Hydraulics and Ministry of Transport Public Works and Water Management, 1994b. SOBEK, technical reference guide. Delft Hydraulics, Delft, the Netherlands.

Department of Irrigation, 1979. Chatra Main Canal Restoration: Note on sediment computer model. Prepared by Sir M MacDonald and Partners for Sunsari Morang Irrigation Project, Ministry of Water Resources, Nepal.

Department of Irrigation, 1982. Sediment control works, two-stage strategy. Prepared by M. Mahmood for Sunsari Morang Irrigation Project, Ministry of Water Resources, Nepal.

Department of Irrigation, 1985a. Completion Report of Koshi river training and Chatra Main Canal Sediment Control Study. Prepared by Coode and Partners for Sunsari Morang Irrigation Project, Ministry of Water Resources, Nepal.

Department of Irrigation, 1985b. Operation and maintenance manual, Vol II: Intake and headreach. Prepared by Coode and Partners for Sunsari Morang Irrigation Project, Ministry of Water Resources, Nepal.

Department of Irrigation, 1987. Design Report Volume I. Prepared by NIPPON KOEI for Sunsari Morang Stage-II Irrigation Project, Ministry of Water Resources, Nepal.

Department of Irrigation, 1990a. Distribution systems canals and canal structures. Design Manual No. M-8 Part 1, Prepared by Sir M MacDonald and Partners for Planning and Design Strengthening Project, Ministry of Water Resources, Nepal.

Department of Irrigation, 1990b. Master plan for irrigation development in Nepal. Main Report, Prepared by Canadian International Water and Energy Consultants for Planning and Design Strengthening Project, Ministry of Water Resources, Nepal.

Department of Irrigation, 1995a. Detail feasibility and design report. Prepared by NIPPON KOEI for Sunsari Morang Irrigation III Project, Ministry of Water Resources, Nepal.

Department of Irrigation, 1995b. Project Operation Plan. Prepared by NIPPON KOEI for Sunsari Morang Irrigation Project, Ministry of Water Resources, Nepal.

Department of Irrigation, 1998. Project Completion Report. Detailed Design and Construction Supervision for Headworks and Desilting Facilities and Water Management, Vol-I, Main Report. Prepared by NEDECO for Sunsari Morang Irrigation Project, Ministry of Water Resources, Nepal.

Department of Irrigation, 2003. Design Report Vol-I, Main Report and Appendices. Prepared by NEDECO for Sunsari Morang Irrigation Project Stage III (phase-I), Ministry of Water Resources, Nepal.

Department of Irrigation, 2007. Historical Background of Irrigation Development in Nepal. Retrieved March 14, 2007 from: http://www.doi.gov.np/developement.php

Depeweg, H. and Mendéz, N., 2007. A New Approach to Sediment Transport in the Design and Operation of Irrigation Canals. Taylor & Francis/Balkema. p. 185, Leiden, the Netherlands.

Depeweg, H. and Paudel, K. P., 2003. Sediment Transport Problems in Nepal Evaluated by the SETRIC Model. Irrigation and Drainage 52: 247-260.

Easter, W., Plusquellec, H. and Subramanian, A., 1998. Irrigation improvement strategy review: A review of bankwide experience based on selected "New Style" projects. Water Resources Thematic Group, World Bank, USA.

Einstein, H., 1950. The bed load function for sediment transportation in open channel flow. Bulletin No. 1026, U.S. Dep. of Agriculture, Washington, USA.

Einstein, H. A., 1942. Formulas for the transportation of bed-load. Trans. ASCE, 107: 133-169.

Einstein, H. A. and Barbarossa, N. L., 1953. River channel roughness. ASCE Transactions, Paper 2528.

Engelund, F., 1966. Hydraulic resistance of alluvial stream. Journal of Hydraulic Division, ASCE, 92(2).

Engelund, F. and Hansen, E., 1967. A monograph on sediment transport in alluvial streams. Teknisk Forlag, Copenhagen, Denmark.

Etcheverry, B. A., 1915. Irrigation Practice and Engineering (Vol. II). McGraw-Hill Book Company, Inc. New York.

FAO, 1988. Irrigation water management: Irrigation methods. Training Manual No. 5, FAO - Food and Agriculture Organization of the UN, Rome.

FAO, 1997. Small-scale irrigation for arid zones: Principles and options. FAO Development Series No. 2, FAO- Food and Agriculture Organization of UN, Rome.

FAO, 1998. Crop evapotranspiration - Guidelines for computing crop water requirements Irrigation and drainage paper No. 56, FAO - Food and Agriculture Organization of the UN, Rome.

FAO, 2003. The irrigation challenge: Increasing irrigation contribution to food security through higher water productivity from canal irrigation systems. Issues Paper No. 4, FAO, Rome, Italy.

FAO, 2006. MASSCOT: a methodology to modernize irrigation services and operation in canal systems. Applications to two systems in Nepal Terai: Sunsari Morang Irrigation System and Narayani Irrigation System. Food and Agriculture Organization of the UN, Rome.

FAO, 2007. Crop prospects and food situations. Report No. 3, Food and Agriculture Organization of UN, Rome.

Fischer, G., Tubiello, F. N., Velthuizen, H. v., et al., 2006. Climate change impacts on irrigation water requirements: Effects of mitigation, 1990-2080.

Technological Forecasting & Social Change, doi:10.1016/j.techfore.2006.05. 021.

Fortier, S. and Scobey, F. C., 1926. Permissible Canal Velocities. Transactions, ASCE, 89: 940-984.

Gailani, J., Zeigler, C. K. and Lick, W., 1991. Transport of suspended solids in the lower Fox River. Journal of Great Lakes Research, 17(4): 479-494.

Galappatti, G. and Vreugdenhil, C. B., 1985. A depth-integrated model for suspended sediment transport. Journal of Hydraulic Research, 23(4): 359-377.

Galappatti, R., 1983. A Depth Integrated Model for Suspended Transport. Report No. 83-7, Delft University of Technology, Delft, the Netherlands.

García-Martínez, R., C., I. S., Power, B. F. d., et al., 1999. A Two-dimensional Computational Model to Simulate Suspended Sediment Transport and Bed Changes. Journal of Hydraulic Research, 37(3).

Ghimire, P. K., 2003. Non-equilibrium Sediment Transport in Irrigation Canals. Unpublished M.Sc. Thesis, IHE, Delft, the Netherlands.

Guo, Q. C. and Jin, Y. C., 1999. Modelling Sediment Transport Using Depth-Averaged and Momentum Equations. Journal of Hydraulic Engineering, ASCE, 125(12).

Guo, Q. C. and Jin, Y. C., 2002. Modeling Nonuniform Suspended Sediment Transport in Alluvial Rivers. Journal of Hydraulic Engineering, ASCE, 128(9).

Hada, G. B., 2003. Nepal ma Yojanabaddha Sinchai Bikaashka Paanch Dashak - Upalabdhi ra Bholika Chunauti (in Nepali). Proc. Golden Jubilee Seminar on Five Decades of Planned Irrigation Development: Achievements and Future Challenges. Kathmandu, Nepal. 134-154.

Halcrow, 2003. ISIS Sediment. User Manual, Halcrow Group Ltd., London, UK.

Halcrow, 2007. Halcrow software: ISIS-Sediment. Retrieved September 23, 2007 from: http://www.halcrow.com/software/solutions/isis-sed.asp

Henderson, F. M., 1966. Open Channel Flow. Macmillan Publishing Co., Inc., New York.

Höfer, A., 1979. The caste hierarchy and the state in Nepal: a study of the Muluki ain of 1854. Khumbu Himal, Bd. 13/2. Innsbruck : Universitätsverlag Wagner.

Holly, F. M. J. and Parrish, J. B., 1992. CanalCAD: Dynamic flow simulation in irrigation canals with automatic gate control. Report No. 196, Iowa Institute of Hydraulic Research, The University of Iowa, Iowa City, Iowa.

Horst, L., 1998. The Dilemmas of Water Division. Considerations and Criteria for Irrigation System Design. IWMI. P. O. Box 2075. Colombo, Sri Lanka.

HR Wallingford, 1988. Chatra Canal Nepal, 1987 Sediment Measurements. Report No. EX 1772, Hydraulic Research Ltd, Wallingford, UK.

HR Wallingford, 1990. Sediment Transport, The Ackers and White Theory Revised. No. SR237, HR Wallingford, England.

HR Wallingford, 1992. DORC: user manual. HR Wallingford, Wallingford, United Kingdom.

HR Wallingford, 2002. SHARC, Sediment and Hydraulic Analysis for Rehabilitation of Canals. Software Manual, HR Wallingford/DFID, UK.

Jansen, P., van Bendegom, L., van den Berg, J., et al. (Eds.), 1979. Principles of river engineering. Pitman press. London.

Kamphuis, J. W., 1974. Determination of sand roughness for fixed beds. Journal of Hydraulic Research, 12(2): 193-202.

Kennedy, R. G., 1895. The prevention of silting in irrigation canals Minutes of the Proceedings, 119(1895): 281-290.

Kerssens, P. J. M., Prins, A. and Rijn, L. C. v., 1979. Model for Suspended Sediment Transport. Journal of Hydraulic Division, ASCE, 105(HY5).

Kerssens, P. J. M. and Van Rijn, L. C., 1977. Model for Non-steady Suspended Sediment Transport. Paper presented at the Seventeenth Congress of the International Association for Hydraulic Research. Baden-Baden.

Khanal, P., 2003. Engineering Participation, The Processes and Outcomes of Irrigation Management Transfer in the Terai of Nepal. Orient Longman private Limited. New Delhi, India.

Klaassen, G. J., 1995. Lane's Balance Revisited. Proc. 6th International Symposium on River Sedimentation. New Delhi, India.

Kouwen, N., 1988. Field estimation of biomechanical properties of grass. Journal of Hydraulic Research, 26(5): 559-568.

Kouwen, N. and Li, R. M., 1980. Biomechanics of vegetated channel linings. Journal of Hydraulic Research, 106(6): 1085-1103.

Krishnamurthy, M. and Christensen, B. A., 1972. Equivalent roughness for shallow channels. Journal of the Hydraulics Division, 98(12): 2257-2263.

Krüger, F., 1988. Flow laws in open channels. Dresden University of Technology, Dresden, Germany.

Lacey, G., 1930. Stable channels in alluvium. Proc. Institutions of Civil Engineers. London, England. Vol. 229, 259-292.

Lane, E. W., 1953. Progress report on studies on the design of stable channels by the Bureau of Reclamation. Proc. ASCE. 79.

Lane, E. W., 1955. Design of stable alluvial channels. Transactions, ASCE, 120: 1234-1260.

Laycock, A., 2007. Irrigation Systems: Design Planning and Construction. CABI. Wallingford, UK.

Lin, B. and Falconer, R. A., 1996. Numerical modelling of three-dimensional suspended sediment for estuarine and coastal waters. Journal of Hydraulic Research, 34(4): 435-456.

Lin, P. N., Huan, J. and Li, X., 1983. Unsteady transport of suspended load at small concentrations. Journal of Hydraulic Engineering, ASCE, 109(1): 86-98.

Lindley, E. S., 1919. Regime Channels. Proc. Punjab Engineering Congress. 7, 63-74.

Lyn, D. A., 1991. Resistance in flat-bed sediment laden flows. Journal of Hydraulic Engineering, ASCE, 117(1): 94-114.

Mahmood, K., 1971. Flow in sand bed channels. Water Resources Technical Report No. 11, Colorado State University, Fort Collins.

Malaterre, P. O., 2007. SIC: Simulation of Irrigation Canals (Ver. 4.26). Retrieved October 16, 2007 from: http://www.canari.free.fr/sic/sicgb.htm

Malaterre, P. O. and Baume, J. P., 1997. SIC 3.0, a simulation model for canal automation design. Paper presented at the International Workshop on the Regulation of Irrigation Canals: State of the Art of Research and Applications, April 22-24, 1997. Marrakech (Morocco).

Mavis, F. T., Liu, T. and Soucek, E., 1937. The Transportation of Detritus by Flowing Water. University of Iowa, Studies in Engineering, USA.

McAnally , W. H., Letter, J. V. and Thomas, W. A., 1986. Two and Three-Dimensional Modelling Systems for Sedimentation. Proc. 3rd International Symposium on River Sedimentation. Jackson, USA.

Méndez, N. V., 1998. Sediment Transport in Irrigation Canals. A.A. Balkema. Rotterdam, the Netherlands.

Meyer, W. B. and Turner II, B. L., 1992. Human Population Growth and Global Land-Use/Cover Change. Annual Review of Ecology and Systematics, 23: 39-61.

Miller, H. P., 1983. Three-Dimensional Free-Surface Suspended Particle Transport in the South Biscayne Bay, Florida. International Journal of Numerical Methods in Fluids, 4: 901-914.

Mishra, S. K., 2004. Operation performance of a modernised irrigation scheme. Case study of Sunsari Morang Irrigation Scheme, Nepal. Unpublished MSc, UNESCO-IHE, Institute for Water Education, Delft, the Netherlands.

Murray-Rust, D. H. and Halsema, G. V., 1998. Effects of construction defects on hydraulic performance of Kalpani distributary, NWFP, Pakistan. Irrigation and Drainage Systems, 12: 323–340.

National Planning Commission, 2007. Nepal's Plans. Retrieved February 17, 2007 from: http://www.npc.gov.np/tenthplan/nepals_plans.htm

Neill, C. R., 1967. Mean Velocity Criterion for Scour of Coarse Uniform Bed Material. Proc. 12th Congress, IAHR. Fort Collins, Colorado. 3, 46-54.

O'Connor, B. A. and Nicholson, J., 1988. A three-dimensional model of suspended particle sediment transport Coastal engineering, 12(2): 157-174.

Olsen, N. R. B., 1999. Two-dimensional numerical modelling of flushing processes in water reservoirs. Journal of Hydraulic Research, 37(1): 3-16.

Orellana, O. M. and Giglio, J. F. V., 2004. Computer aided design of canals. Unpublished MSc Thesis, UNESCO-IHE, Delft, the Netherlands.

Parajuli, U. N. and Sharma, K. R., 2003. Golden Jubilee Seminar on Five Decades of Planned Irrigation Development: Achievements and Future Challenges. Kathmandu, Nepal.

Paudel, K. P., 2002. Evaluation of the Sediment Transport Model SETRIC. Unpublished M.Sc. Thesis, International Institute for Infrastructure Hydraulic and Environmental Engineering, IHE, Delft, the Netherlands.

Pereira, L. S., Gilley, J. R. and Jensen, M. E., 1996. Research Agenda on Sustainability of Irrigated Agriculture. Journal of Irrigation and Drainage Engineering, 122(3): 172-177.

Pesala, B., 2006. The Dhammapada and Commentary. Retrieved April 05, 2007 from: http://aimwell.org/assets/Dhammapada.pdf

Petryk, S. and Bosmajian, G., 1975. Analysis of flow through vegetation. Journal of Hydraulic Division, ASCE, 101(7): 871-884.

Pluesquellec, H., 2002. How Design, Management and Policy Affect the Performance of Irrigation Project, Emerging Modernization Procedure and Design Standards. FAO, Regional Office for Asia and Pacific, Bangkok, Thailand.

Poudel, S. N., 2003. Nepal Ma Sinchai: Jaljanya Sanstha Ra Sansthagat Vikas (in Nepali). Jalshrot Vikas Sanstha. Kathmandu, Nepal.

Pradhan, T. M. S., 1996. Gated or ungated: water control in government-built irrigation systems: comparative research in Nepal. Wageningen University, Wageningen.

Querner, E. P., 1997. A model to estimate timing of aquatic weed control in drainage canals. Part 1. Irrigation and Drainage Systems 11: 157-169.

Rajaratnam, N. and Subramanya, K., 1967. Flow equation for the sluice gate. Journal of Irrigation and Drainage Division, ASCE, 93(3): 167–186.

Ranga Raju, K. G., 1981. Flow through open channels. Tata McGraw-Hill. New Delhi, India.

Ranga Raju, K. G., Dhandapani, K. R. and Kondup, D. M., 1977. Effect of Sediment Load on Stable Sand Canal Dimensions. Journal of Waterway, Port, Coastal and Ocean Division, ASCE, 103(WW2): 241-249.

Raudkivi, A. J., 1990. Loose Boundary Hydraulics (3rd Edition ed.). Pergamon Press. Great Britain.

Regmi, D. R., 1969. Ancient Nepal. Firma K. L. Mukhopadhyay. p. 364, Calcutta.

Renault, D., 2003. Irrigation System Operation, Diagnosis, Performance and Design. Training Workshop on Modernization of Irrigation Systems for Integrated Water Resource Management. Nepal. Land and Water Dev Division, FAO, Rome, Italy.

Ribberink, J. S., 1986. Integration to a depth-integrated model for suspended transport (Galappatti, 1983). No. 6-86, TU Delft, Faculty of Civil Engineering, Delft, the Netherlands.

Ribberink, J. S., 1987. Mathematical modelling of one-dimensional morphological changes in rivers with non-uniform sediment. No. 169-6548, TU Delft, Faculty of Civil Engineering, Delft, the Netherlands.

Rooseboom, A. and Annandale, G. W., 1981. Techniques applied in determining sediment loads in South African rivers. In: Erosion and Sediment Transport Measurement. Proc. Florence Symposium, IAHS. 133, 219-224.

Schultz, B., 2002. Lecture Notes on Land and Water Development. IHE. Delft, the Netherlands.

Schultz, B., Thatte, C. D. and Labhsetwar, V. K., 2005. Irrigation and drainage: Main contributors to global food productivity. Irrigation and Drainage, 54: 263–278.

Schultz, B. and Wrachien, D. D., 2002. Irrigation and Drainage Systems: Research and Development in the 21st Century. Irrigation and Drainage, 51: 311-327.

Schuurmans, J., Clemmens, A. J., Dijkstra, S., et al., 1999. Modelling of Irrigation and Drainage Canals for Controller Design. Journal of Irrigation and Drainage Engineering, ASCE, 125(6).

Shah, S. G. and Singh, G. N., 2001. Irrigation Development in Nepal Investment, Efficiency and Institution. Research Report Series No. 47, Winrock International, Kathmandu, Nepal.

Shen, H. W., 1976. Stochastic approaches to water resources (Vol. I and II). Colorado State university. Fort Collins, Colorado.

Sheng, Y. P. and Butler, H. L., 1982. Modeling coastal currents and sediment transport. Proc. 18th International Conference on Coastal Engineering. Cape Town, South Africa.

Sherpa, K., 2005. Use of sediment transport model SETRIC in an irrigation canal. Unpublished MSc Thesis, UNESCO-IHE, Delft, the Netherlands.

Shields, A., 1936. Anwendung der Aehnlichkeits-Mechanik und der Turbulenzforschung auf die Geschiebebewegung. Preussische Versuchsanstalt für Wasserbau und Schiffbau, Berlin, Heft 26.

Shoemaker, H. J., 1983. Dynamica van Morphologische Processen (in Dutch). Delft University of Technology, Faculty of Civil Engineering, Delft, the Netherlands.

Simons, D. and Senturk, F., 1992. Sediment Transport Technology, Water and Sediment Dynamics. Water Resources Publications. Colorado, USA.

Simons, D. B. and Albertson, M. L., 1963. Uniform Water Conveyance Channels in Alluvial Material. Transactions, ASCE, 128(1): 65-167.

Sloff, K., 2006. Morphological modelling using SOBEK-RE. WL | Delft Hydraulics, Delft, the Netherlands.

Smith, T. J. and O'Connor, B. A., 1977. A Two-Dimensional Model for Suspended Sediment Transport. Proc. IAHR-Congress. Baden-Baden, West Germany.

Spaans, W., 2000. Duflow Modelling Studio, User's Guide. Lecture Note, International Institute for Infrastructural, Hydraulic and Environmental Engineering, Delft, the Netherlands.

Stevens, M. A. and Nordin, C. F., 1987. Critique of the Regime Theory for Alluvial Channels. Journal of Hydraulic Engineering, ASCE, 113(11): 1359-1380.

Task Committee on Preparation of Sedimentation Manual, 1966a. Sediment transportation mechanics: Initiation of motion. Journal of Hydraulic Division, ASCE, 92(2): 291-314.

USACE, 2006. River Analysis System, HEC-RAS. Release Notes, US Army Corps of Engineers, Hydrologic Engineering Center, USA.

Van Rijn, L. C., 1982. Equivalent roughness of alluvial bed. Journal of the Hydraulics Division, 108(10): 1215-1218.

Van Rijn, L. C., 1984a. Sediment Transport Part I: Bed load Transport. Journal of Hydraulic Division, ASCE, 110(10).

Van Rijn, L. C., 1984b. Sediment Transport Part II: Suspended Load Transport. Journal of Hydraulic Division, ASCE, 110(11).

Van Rijn, L. C., 1984c. Sediment Transport part III: Bed forms and alluvial roughness. Journal of Hydraulic Division, ASCE, 110(12): 1733-1754.

Van Rijn, L. C., 1986. Mathematical Modelling of Suspended Sediment in Non-Uniform Flows. Journal of Hydraulic Engineering, ASCE, 112(6): 433-455.

Van Rijn, L. C., 1987. Mathematical modelling of morphological processes in the case of suspended sediment transport. Delft Hydraulics Communication No. 382, Delft, the Netherlands.

Van Rijn, L. C., 1993. Principles of Sediment Transport in Rivers, Estuaries and Coastal Areas. Aqua Publications. Amsterdam, the Netherlands.

Van Rijn, L. C. and Meijer, K., 1988. Three-Dimensional Mathematical Modelling of Suspended Sediment Transport in Currents and Waves. Proc. IAHR Symposium. Copenhagen, Denmark.

Vanoni, V. A., 1975. Sedimentation engineering. ASCE manuals and reports on engineering practice. American Society of Civil Engineers, USA.

Vanoni, V. A. and Brooks, N. H., 1957. Laboratory Studies of the Roughness and Suspended Load of Alluvial Streams. No. E-68, California Institute of Technology, California, USA.

Varshney, R. S., Gupta, S. C. and Gupta, R. L., 1992. Theory and design of irrigation structures, Volume-I, Channels and Tubewells (6th ed.). Nem Chand and Brothers. Roorkee 247667, India.

Vlugter, H., 1962. Sediment transportation by running water and design of stable channels in alluvial soils (in Dutch). De Ingenieur, The Hague, the Netherlands.

Vreugdenhil, C., 1989. Computational Hydraulics. Springer-Verlag. Berlin, Germany.

Vreugdenhil, C. B. and Vries, M. de, 1967. Computations on Non-steady Bedload-transport by a Pseudo-viscosity Method. Hydraulic Laboratory Delft, the Netherlands.

Vreugdenhil, C. B. and Wijbenga, J. H. A., 1982. Computation of Flow Patterns in Rivers. Journal of Hydraulic Division, ASCE, 108(11): 1296-1310.

Wahl, T. L., Clemmens, A. J., Bos, M. G., et al., 2001. Winflume: Windows-Based Software for the Design of Long-Throated Measuring Flumes: U.S. Bureau of Reclamation, Denver, Colorado.

Wahl, T. L., Clemmens, A. J., Replogle, J. A., *et al.*, 2005. Simplified design of flumes and weirs. Irrigation and Drainage, 54: 231-247.

Wang, S. Y. and Adeff, S. E., 1986. Three-dimensional modeling of river sedimentation processes. Proc. 3rd International Symposium on River Sedimentation. University of Mississippi, USA. 1594-1601.

Wang S. Q., White, W. R. and Bettess, R., 1986. A rational approach to river regime. Proc. 3rd International Symposium on River Sedimentation. Jackson, Mississippi, USA.

Wang, Z. B., 1989. Mathematical Modelling of Morphological Processes in Estuaries. Delft University of Technology, Delft, the Netherlands.

Wang, Z. B. and Ribberink, J. S., 1986. The validity of a depth-integrated model for suspended sediment transport. Journal of Hydraulic Research, 24(1).

Water and Energy Commission Secretariat, 1982. Performance Review of Public Sector Intensive Irrigation Based Agriculture Development Projects. Report No. 3/2/1407821/3, WECS/IBRD, His Majesty's Government, Nepal.

Water and Energy Commission Secretariat, 2003. Water Resource Strategy Nepal. Water and Energy Commission Secretariat Kathmandu, Nepal.

White, W. R., Bettess, R. and Paris, E., 1982. Analytical Approach to River Regime. Journal of Hydraulic Division, ASCE, 108(10): 1179-1193.

White, W. R., Paris, E. and Bettess, R., 1980. The frictional characteristics of alluvial stream; A new approach Proceedings of the Institution of Civil Engineers, 69(2): 737-750.

White, W. R., Paris, E. and Bettess, R., 1981b. Tables for the design of stable alluvial channels. No. IT 208, Hydraulics Research Station, Wallingford, UK.

Woo, H. and Yu, K., 2001. Reassessment of Selected Sediment Discharge formulas. Proc. XXIX IAHR Congress. Beijing, China. 224-230.

World Bank, 1997. Project Appraisal Document, Nepal. Nepal Irrigation Sector Project No. Report No. 17104-NEP, Rural Development Sector Unit, South Asia Region, World Bank,

World Bank, 2005. World development report 2006, Equity and Development. The International Bank for Reconstruction and Development / The World Bank, Washington DC 20433.

Wu, F. C., Shen, H. W. and Chou, Y. J., 1999. Variation of roughness coefficient for un-submerged and submerged vegetation. Journal of Hydraulic Engineering, ASCE, 125(9): 934-942.

Yalin, M. S., 1977. Mechanics of Sediment Transport. Pergamon Press. Oxford, Great Britain.

Yang, C. T., 1972. Unit Stream Power and Sediment Transport. Journal of Hydraulic Division, ASCE, 98(HY10): 1805-1826.

Yang, C. T., 1973. Incipient Motion and Sediment Transport. Journal of Hydraulic Division, ASCE, 99(HY10): 1679-1704.

Yang, C. T. and Molinas, A., 1982. Sediment Transport and Unit Stream Power Function. Journal of the Hydraulics Division, 108(6): 774-793.

Yang, C. T. and Song, C. C. S., 1979. Theory of Minimum Rate of Energy Dissipation. Journal of Hydraulic Division, ASCE, 105(HY7): 769-784.

Yang, C. T. and Wan, S., 1991. Comparisons of Selected Bed-Material Load Formulas. Journal of Hydraulic Engineering, 117(8): 973-989.

Yang, C. T., Yong, C. C. S. and Woldenberg, M. J., 1981. Hydraulic Geometry and Minimum Rate of Energy Dissipation. Water Resources Research, 17(4): 1014-1018.

Yang, S. Q. and Lim, S. Y., 1997. Mechanism of energy transportation and turbulent flow in a 3D channel. Journal of Hydraulic Engineering, ASCE, 123(8): 684-692.

Yang, S. Q., Yu, J. X. and Wang, Y. Z., 2004. Estimation of diffusion coefficients, lateral shear stress and velocity in open channels with complex geometry. Water Resources Research, 40(W05202).

Yen, B. C., 2002. Open Channel Flow Resistance. Journal of Hydraulic Engineering, ASCE, 128(1): 20-39.

Ziegler, C. K. and Lick, W., 1986. A numerical model of the re-suspension, deposition and transport of fine-grained sediments in shallow water. USCB Report No. ME-86-3, University of California, Santa Barbara, California, USA.

Ziegler, C. K. and Nisbet, B., 1994. Fine-grained sediment transport in Pawtuxet river, Rhode Island. Journal of Hydraulic Division, ASCE, 120(5): 561-576.

Appendix A: Flow diagrams

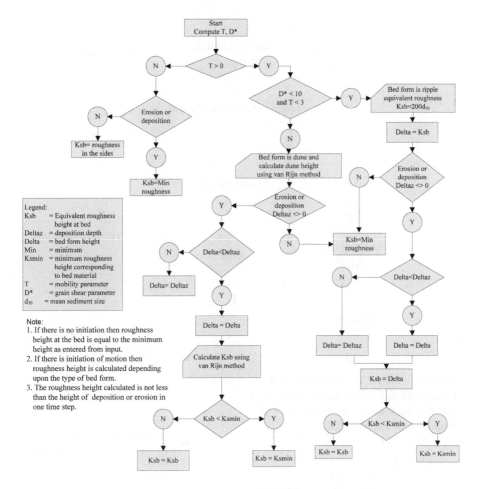

Figure A.1 Roughness prediction procedure depending upon the flow condition and sediment parameter.

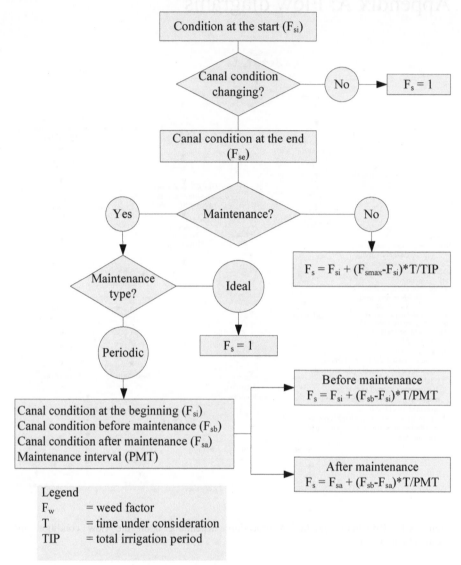

Figure A.2   Roughness adjustment procedure for the changing canal condition with or without maintenance scenarios.

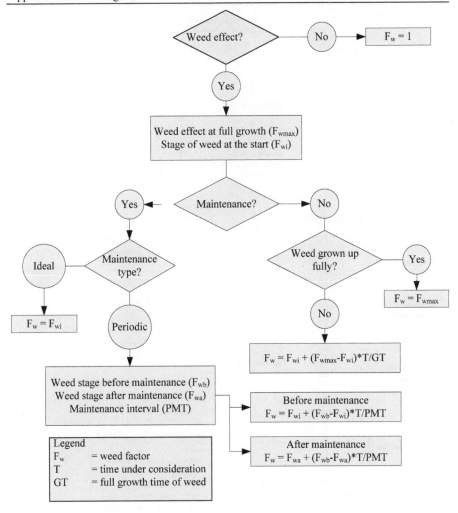

Figure A.3   Weed height calculation method.

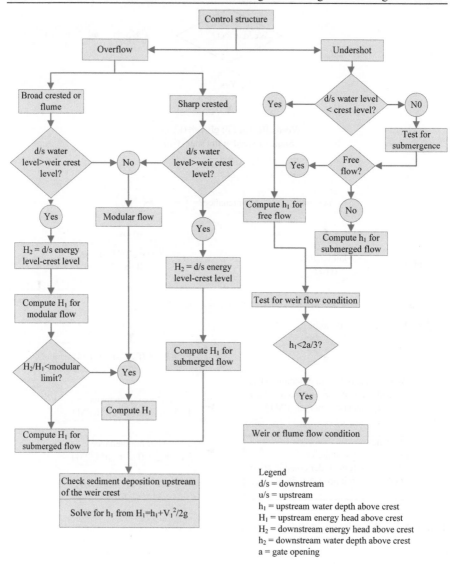

Figure A.4   Calculation steps for water depth upstream of a flow control structure.

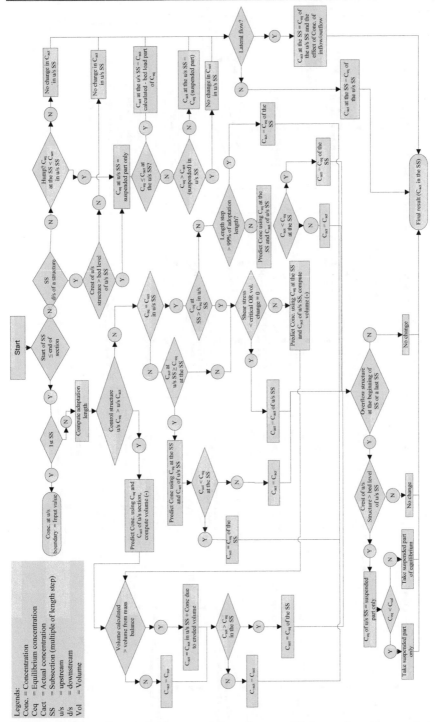

Figure A.5   Flow diagram for the computation of non-equilibrium sediment transport at any section.

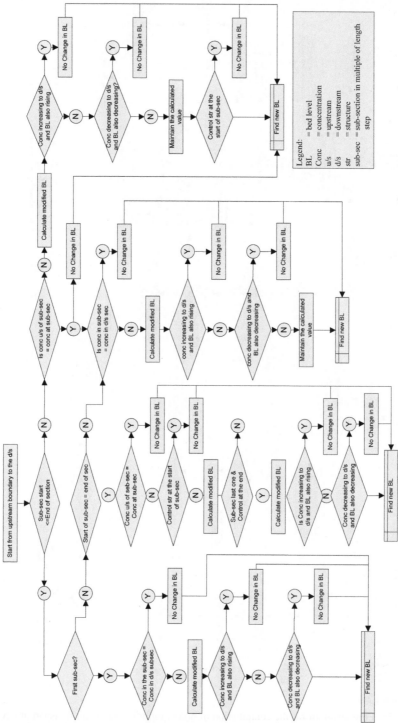

Figure A.6   Flow diagram for the bed level change (solution of sediment mass balance equation).

Appendix B: Water and sediment data

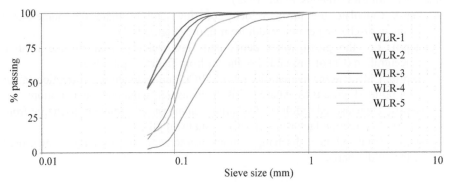

Figure B.1   Gradation curve of deposited sediment samples upstream of water level regulators (WLR) in Secondary Canal S9 (2004).

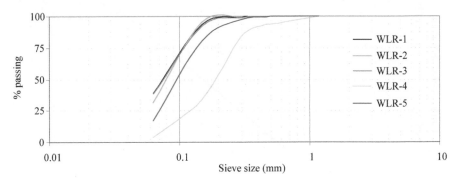

Figure B.2   Gradation curve of deposited sediment samples upstream of water level regulators (WLR) in Secondary Canal S9 (2005).

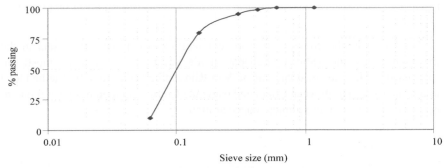

Figure B.3 Gradation curve of suspended sediment samples collected near the head of Secondary Canal S9 during 2005 (size > 63 μm only).

Determination of total sediment load from measured suspended load

The bed load part of the total sediment concentration has been determined using Colby's (1957) method. Stepwise calculation for the determination of total sediment loads is as follows (Simons, *et al.*, 1992).

- step-i. Calculate the mean velocity of flow and estimate the unmeasured sediment discharge per unit width from figure (Figure B.4);
- step-ii. For the given water depth and flow velocity estimate the relative concentration (Cr) of suspended sediment from figure (Figure B.5);
- step-iii. Calculate the availability ratio Cs'/Cr, and estimate the correction factor from figure (Figure B.6), where Cs' is the measured sediment concentration;
- step-iv. Unmeasured sediment discharge = unmeasured sediment discharge from step (i) * correction factor in step (iii) * canal width B;
- step-v. Total sediment discharge = unmeasured sediment discharge + measured sediment discharge.

The computed unmeasured sediment discharge for bed load part was then added to the measured part to get the total concentration.

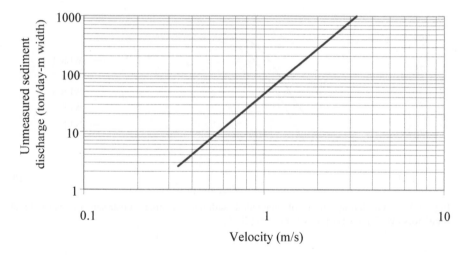

Figure B.4   Unmeasured sediment discharge for flow velocity adapted from Colby (1957).

For the hydraulic conditions mean sediment size (d_{50}) generally found in the irrigation canals the sediment transport mode has been found to be mostly in suspended mode when evaluated using Van Rijn method (Figure 6.2). The bed load in Secondary Canal S9 when calculated using the Colby's method has been found to be 20 to 30% of the total sediment concentration.

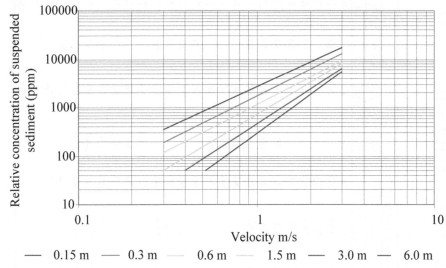

Figure B.5   Relative concentration for different depth and mean velocity adapted from Colby (1957).

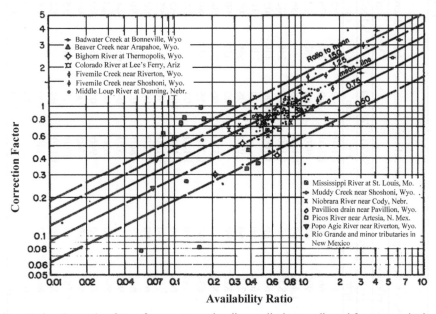

Figure B.6   Correction factor for unmeasured sediment discharge adjusted for mean velocity with a measure of availability of sands after Colby (1957) (from Simons and Senturk, 1998).

Role of sediment for the design and management of irrigation canals

Table B.1   Water inflow to Secondary Canal S9 and water delivery to sub-secondary and tertiary canals (2004).

Date	Water inflow	Sediment inflow		Water outflow rates and set points													
		> 63 μm	Total	SS9A	SS9C	SS9B	WLR-1	SS9D	WLR-2	S9-T2	S9-T1	WLR-3	SS9E	WLR-4	SS9F	WLR-5
	l/s	ppm	ppm	l/s	l/s	l/s	cm	l/s	cm	l/s	l/s	cm	l/s	cm	l/s	cm
26-Jul	4,840	180	680	110		1,140	100	240	65	170	220	66		65		60
27-Jul	5,840	110	660	110		1,140	100	280	76	170	220	80		80	40	70
28-Jul	5,840	230	990	60		980	92	480	72	170	220	75		80		78
29-Jul	6,140	180	1,490	160		1,140	108		86	150	450	85		90	50	80
30-Jul	6,140	230	850	160		1,140	108	240	82	170	410	75		90	40	78
31-Jul	4,170	130	650	160		1,100	100	440	105	150	310	82	780	90	40	75
1-Aug	4,170	120	660	160			100	730	110	250	590	112	660	90	40	74
2-Aug	4,840	300	260	160		980	108	680	105	250	590	108	690	90	40	74
3-Aug	5,840	240	830	190			100	630	90			83	850	90		80
4-Aug	6,140	190	840			370	100	240	88			86	1,250	90		81
5-Aug	6,140	110	830	160		1,260	115	440	80			79	850	80		74
6-Aug	5,550	130	740	160		1,140	104		80	220	590	90		76	50	70
7-Aug	5,990	200	640	160		1,140	100		85	270	450	90		85	40	75
8-Aug	6,140	120	510	160		1,140	96		85	200	520	90		87	50	75
9-Aug	5,990	170	690	160		220	90		90	200	520	95		90	50	75
10-Aug	6,140	50	600			860	102	730	90			75	1,010	84		71
11-Aug	5,990	60	980			520	104	630	84			72	850	78		74
12-Aug	6,140	60	890	190		650	100	630	84			72	1,010	84		70
13-Aug	4,840	60	820	160		940	105	630	90	150	310	80	1,110	95	50	75
14-Aug	5,990	120	800	160		760	100	550	90	170	410	84	1,040	95	50	76
15-Aug	5,120	420	1,580	160		760	95	440	85	210	450	94	850	90	40	70
16-Aug	6,140	220	830	140		1,140	94		84	170	520	80		87	40	72
17-Aug	4,700	170	880	220		1,140	100		68			67		78		74
18-Aug	6,290	220	1,010				110		100			84		92		79
19-Aug	6,140	210	1,270				110		100			83		90		80

Appendix B: Water and sediment data

Date	Water inflow	Sediment inflow > 63 µm	Total	SS9A	SS9C	SS9B	WLR-1	SS9D	WLR-2	S9-T2	S9-T1	WLR-3	SS9E	WLR-4	SS9F	WLR-5
	l/s	ppm	ppm	l/s	l/s	l/s	cm	l/s	cm	l/s	l/s	cm	l/s	cm	l/s	cm
20-Aug	5,990	210	1,140				110		100			84		90		78
21-Aug	5,990	300	1,270				110		100			84		92		78
22-Aug	6,290	330	1,110	220		1,220	100		85	210	500	74		78	40	75
23-Aug	7,060	200	750	140		1,260	105		89	210	520	75		84	40	75
24-Aug	7,060	140	790	160		1,260	105		89	210	520	75		93	50	80
25-Aug	7,060	240	950	110		1,260	105		89		520	85		95	50	80
26-Aug	6,750	190	850			860	95		92		100	88		106	30	94
27-Aug	6,140	310	1,030				98		89			88		105		100
28-Aug	4,170	170	960				104		66			65		80		74
29-Aug	4,170	350	980				106		66			65		73		68
30-Aug	5,990	310	1,190				97		90			88		90		86
31-Aug	6,290	220	880				99		91			89		90		87
2-Sep	6,290	250	1,080	160		1,260	110		84	100	260	79		92	40	75
3-Sep	5,550	240	1,300	60		1,060	105		78	70	220	72		88	40	70
4-Sep	6,290	260	1,670	160		1,260	112		84	100	410	90		88	40	71
5-Sep	5,990	240	1,690	160		1,260	105		81	90	220	90		88	40	70
6-Sep	6,140	210	1,080	60		1,060	95		88	100	260	83		95	50	77
7-Sep	6,140	120	490	60		1,140	100		89	100	310	83		95	50	76
9-Sep	4,170	440	1,180			520	100		63			62		72		70
10-Sep	4,170	490	1,510			520	102		63			62		72		70
11-Sep	3,910	240	590				100		67			64		77		73
12-Sep	3,530	100	490				90		86			60		70		68
13-Sep	4,170	70	740				100		88			60		77		72
14-Sep	4,170	60	320				100		88			60		77		72
15-Sep	5,550	140	410	190		1,390	115		102			60		77		72
16-Sep	7,060	370	1,970	230		1,300	110		92	220	370	93		90	30	82
17-Sep	7,380	40	440	230		1,140	112		97	300	520	95		95	40	85
18-Sep	7,380	290	760	190		1,220	98		94	270	520	95		90	50	85

Role of sediment in the design and management of irrigation canals

Date	Water inflow	Sediment inflow		Water outflow rates and set points													
		>63 μm	Total	SS9A	SS9C	SS9B	WLR-1	SS9D	WLR-2	S9-T2	S9-T1	WLR-3	SS9E	WLR-4	SS9F	WLR-5
	l/s	ppm	ppm	l/s	l/s	l/s	cm	l/s	cm	l/s	l/s	cm	l/s	cm	l/s	cm
19-Sep	7,870	150	600	190		1,260	110		99	270	520	95		105	50	85
20-Sep	7,870	260	1,380	190		1,390	106		94	300	590	92		97	50	78
21-Sep	7,870	120	1,160	190		1,390	108		93	300	590	92		97	50	77
23-Sep	6,290	150	550				99		100			86		90		77
24-Sep	6,140	370	660				97		100			85		90		78
25-Sep	6,440	100	330				100		90			91		100		85
26-Sep	7,060	160	1,020				105		98			94		105		90
27-Sep	7,060	280	600				105		98			96		105		91
28-Sep	7,060	120	900				105		99			97		105		93
30-Sep	7,060	220	540	160		1,260	110		95	150	370	90		105	50	84
1-Oct	6,290	940	1,140	160		1,260	108		87	150	500	88		92	40	73
2-Oct	6,290	930	950	160		1,260	108		87	150	520	87		91	40	73
3-Oct	6,290	840	1,160	160		1,060	106		87	300	520	87		90	50	85
4-Oct	6,290	1,550	1,680	160		1,060	105		87	270	520	87		89	50	85
5-Oct	6,140	750	1,050	60		1,020	103		86	300	590	87		88	50	85
7-Oct	2,940	390	950				110		98			48		65		45
8-Oct	2,940	230	590				110		98			48		60		45
9-Oct	2,380	230	480				108		80			44		70		33
10-Oct	2,380	150	420				108		75			43		80		70
11-Oct	6,290	180	570				99		95			94		98		84
12-Oct	5,550	200	380			620	92		86		260	82		86	40	69
13-Oct	5,550	210	430			690	92		86		220	82		87	40	72
14-Oct	5,550	90	400			690	85		82		260	78		90	30	65
15-Oct	5,550	90	400				84		96	150	220	74		78	30	63
16-Oct	5,550	90	400				84		80	210	370	79		83	30	60
17-Oct	5,550	20	140			370	88		84	250	370	78		77	30	62
18-Oct	5,840	20	180			520	90		86		190	82		87	30	72
19-Oct	5,840	10	110			860	110		80		150	79		84	30	72

Appendix B: Water and sediment data

Date	Water inflow	Sediment inflow		Water outflow rates and set points												
		>63 μm	Total	SS9A	SS9C	SS9B	WLR-1	SS9D	WLR-2	S9-T2	S9-T1	WLR-3	SS9E	WLR-4	SS9F	WLR-5
	l/s	ppm	ppm	l/s	l/s	l/s	cm	l/s	cm	l/s	l/s	cm	l/s	cm	l/s	cm
20-Oct	6,290	40	170	140		860	110		85	50	80	83		87	30	74
21-Oct	6,290	20	110	140		1,060	110		82	100	300	100		83		77
22-Oct	6,290	300	310	140		1,140	110		82	20	30	80		88		82
23-Oct	6,290	10	100			1,140	108		81	20	30	80		88		82

Table B.2   Water inflow to Secondary Canal S9 and water delivery to sub-secondary and tertiary canals (2005).

Date	Water inflow	Sediment inflow		Water outflow and set points												
		>63 μm	Total	SS9A	SS9C	SS9B	WLR-1	SS9D	WLR-2	S9-T2	S9-T1	WLR-3	SS9E	WLR-4	SS9F	WLR-5
	l/s	ppm	ppm	l/s	l/s	l/s	cm	l/s	cm	l/s	l/s	cm	l/s	cm	l/s	cm
8-Jul	4,840	220	1,280		220	980	112	280	70	90	500	90		69	1,010	94
9-Jul	4,840	1,620	2,180		220	860	112	280	70	170	450	90		72	980	94
10-Jul	5,550	560	1,440		260		115	630	90			90	690	94		88
11-Jul	5,550	320	770		310		112	680	87			90	1,010	100		84
12-Jul	4,840	310	1,720		410		108	680	80			80	850	100		70
13-Jul	4,840	550	1,110		410		108	680	82			80	780	102		72
14-Jul	4,840	550	1,400			1,260	110		70	170	370	90		75	850	85
15-Jul	4,840	480	970			1,260	108		70	170	410	90		78	910	85
16-Jul	4,840	1,230	1,720			1,260	108		75	170	450	90		78	550	65
17-Jul	4,840	810	1,240			1,220	110		75	20	100	80		82	850	70
18-Jul	4,840	1,040	1,760		370		105	160	86			80	980	84	910	84
19-Jul	4,840	780	1,380		370		108	160	86			80	690	84	850	75
20-Jul	4,980	830	2,200		410		108	440	82		370	100	550	82	610	75
21-Jul	4,430	1,310	2,810		220		100	440	76		190	70	500	80	660	62

Role of sediment in the design and management of irrigation canals

Date	Water inflow	Sediment inflow		Water outflow and set points												
		> 63 µm	Total	SS9A	SS9C	SS9B	WLR-1	SS9D	WLR-2	S9-T2	S9-T1	WLR-3	SS9E	WLR-4	SS9F	WLR-5
	l/s	ppm	ppm	l/s	l/s	l/s	cm	l/s	cm	l/s	l/s	cm	l/s	cm	l/s	cm
22-Jul	4,700	560	1,550			1,260	105		72		100	70		82	1,010	105
23-Jul	4,840	310	1,420			1,260	105		72	170	310	80		80	1,010	105
24-Jul	4,840	650	1,570			1,260	105		70	100	370	80		82	910	92
25-Jul	4,840	580	1,290			1,260	104		72	100	310	90		80	850	95
26-Jul	4,840	430	940		410		105		86			80	690	110	500	95
27-Jul	4,840	390	1,120		410		115		105			80	780	115	500	97
28-Jul	4,840	340	1,030		410		112		105			90		120		78
29-Jul	4,840	270	790		450		115		105			80		120		78
30-Jul	4,840	410	1,200			1,060	110		76			100		82		80
31-Jul	6,290	260	760			1,220	115		85			100		92		85
1-Aug	6,290	570	1,120			1,220	115		85			100		95		92
2-Aug	6,290	600	1,080			1,220	115		84			100		94		90
3-Aug	6,290	190	660		450		120		120			100		106		95
4-Aug	6,290	700	1,240	160	370		120		120			90		106		92
5-Aug	6,290	380	970	160	410		120		120			90		106		92
6-Aug	6,290	430	1,080	160	410		120		90			90		110		90
7-Aug	4,170	220	720			860	95		70			100		85		82
8-Aug	2,380	300	860			240	55		58			50		60		68
9-Aug	4,840	50	1,430	160		1,260	95		72			100		80		72
10-Aug	6,140	270	1,340			1,140	115		95			100		108		105
11-Aug	6,140	450	1,080		520		120		95			100		110		102
12-Aug	3,170	600	1,270		220		92		65			70		82		75
13-Aug	4,700	1,200	2,630		220		85		90			90		95		98
14-Aug	4,840	500	1,120		260		88		86			90		98		98
15-Aug	4,170	660	1,010			1,060	90		65			70		80		75

Appendix B: Water and sediment data

Date	Water inflow	Sediment inflow		Water outflow and set points													
		>63 µm	Total	SS9A	SS9C	SS9B	WLR-1	SS9D	WLR-2	S9-T2	S9-T1	WLR-3	SS9E	WLR-4	SS9F	WLR-5
	l/s	ppm	ppm	l/s	l/s	l/s	cm	l/s	cm	l/s	l/s	cm	l/s	cm	l/s	cm
16-Aug	3,530	320	1,240			520	75		65			90		80		75
17-Aug	4,840	320	1,730			860	75		80			100		98		88
18-Aug	5,840	140	830			1,060	75		89			90		100		100
19-Aug	5,840	320	950		450		105		95			100		108		105
20-Aug	5,840	250	870		450		105		96			100		108		102
21-Aug	5,550	430	1,090		410		100		95			90		112		105
22-Aug	5,840	530	1,550		410		105	780	87	150	260	110	690	115	500	94
23-Aug	5,840	570	1,260	160		1,260	105		87	150	500	120		86	780	92
24-Aug	5,840	180	740	160		1,260	105		86	70	520	120		90	850	95
25-Aug	4,840	600	1,210	160		760	80		82	50	410	110		85	780	95
26-Aug	2,380	450	1,160			130	55		56	50	60	50		65	690	92
27-Aug	2,710	220	1,310		80		78	30	60			60	350	63	420	95
28-Aug	2,940	270	1,440		150		80	30	60			60	350	65	390	92
29-Aug	2,940	240	1,250		220		90	160	62			60	350	63		58
30-Aug	2,940	340	990		220		90	110	60			60	350	63		58
31-Aug	4,840	260	1,320	160		1,220	108		72	70	190	90		75	610	90
1-Sep	4,840	710	1,310	160		1,060	105		76	150	500	120		75	910	100
2-Sep	6,290	800	1,300	160		1,390	120		90	150	500	120		95	1,010	105
3-Sep	6,290	410	930	160		1,220	110		90	150	410	110		98	1,010	105
4-Sep	6,290	610	1,400		520		110	810	90			90	1,010	118		85
5-Sep	6,290	850	1,330		450		108	780	94			90	910	100	850	100
6-Sep	5,550	610	1,190		520		108	730	88			80	690	88		82
7-Sep	5,550	690	1,210		520		115	810	84	210		80	690	86		80
8-Sep	5,550	870	1,720	160		1,140	115		84	250	520	110		85	850	98
9-Sep	5,550	520	1,320	160		1,260	110		84	210	410	110		85	850	98

Role of sediment in the design and management of irrigation canals

Date	Water inflow	Sediment inflow >63 μm	Sediment inflow Total	SS9A	SS9C	SS9B	WLR-1	SS9D	WLR-2	S9-T2	S9-T1	WLR-3	SS9E	WLR-4	SS9F	WLR-5
	l/s	ppm	ppm	l/s	l/s	l/s	cm	l/s	cm	l/s	l/s	cm	l/s	cm	l/s	cm
10-Sep	5,550	610	1,310	160		1,260	110		82	250	450	110		85	910	100
11-Sep	6,290	110	2,990	160		1,260	110		86	250	590	110		95	850	98
13-Sep	4,170	470	730		310	1,060	80	480	86			60	910	100		45
14-Sep	6,290	780	1,210		310	860	90	680	105			90	910	100		80
15-Sep	6,290	680	1,380		410	860	100	680	105			90	910	112		78
16-Sep	6,290	580	1,150		450	1,220	108	630	95			80	850	100		75
17-Sep	6,290	620	1,270		220	1,260	108		86	250	410	120	850	108		95
18-Sep	6,140	450	1,080		310	1,260	108		84	250	450	120		92	850	100
19-Sep	6,290	610	1,040			1,260	100		85	270	500	110		96	1,010	100
20-Sep	6,290	370	620	160		1,390	115		85	270	500	110		96	1,010	100
21-Sep	6,140	710	1,920		520		110	780	112	150	450	110	850	105	1,010	78
22-Sep	6,290	630	1,790		450		115	810	115			90	1,010	118		82
23-Sep	6,290	820	1,400		450		112	810	115			90	1,010	118		82
24-Sep	6,290	960	1,610		520		115	810	115			90	1,110	112		82
25-Sep	5,550	1,030	1,730			1,260	100		80	270	520	110		85	1,010	110
26-Sep	5,550	740	1,410			1,260	100		80	250	450	110		88	980	100
27-Sep	5,550	630	1,090			1,480	102		80	210	450	110		88	980	100
28-Sep	6,290	1,300	2,740			1,390	92		90	270	630	120		92	1,010	110
29-Sep	7,060	1,260	1,730		410		90	780	102			100	1,110	105	1,010	110
30-Sep	7,060	1,090	1,690		410		92	780	102			100	1,110	105	1,010	110
1-Oct	7,060	1,330	1,920		410		85	780	102			100	1,110	105	1,010	110
2-Oct	7,060	1,330	2,000		410		75	780	100		370	100	1,010	102		90
3-Oct	7,060	1,310	2,090	160	190	1,480	112	200	102		370	90	1,010	95		78
4-Oct	7,060	830	1,850		190	1,480	110	160	90	210		90	1,010	95		90
5-Oct	7,060	1,160	2,010		220	1,480	108	240	90	210		90	1,110	95		82

Appendix B: Water and sediment data

Date	Water inflow	Sediment inflow		Water outflow and set points												
		> 63 μm	Total	SS9A	SS9C	SS9B	WLR-1	SS9D	WLR-2	S9-T2	S9-T1	WLR-3	SS9E	WLR-4	SS9F	WLR-5
	l/s	ppm	ppm	l/s	l/s	l/s	cm	l/s	cm	l/s	l/s	cm	l/s	cm	l/s	cm
6-Oct	7,060	1,160	2,440			1,480	70	240	90	210	590	110		100	1,250	115
7-Oct	7,060	1,010	1,640			1,480	68	330	90	210	450	110		100	1,250	115
8-Oct	6,290	1,250	1,880			1,480	82	280	85	170	310	100		95	980	105
9-Oct	6,290	1,120	1,620			1,390	80	280	85	170	310	100		95	980	105
10-Oct	7,060	590	870		310	980	102	280	96	250	520	110		95	1,010	110
11-Oct	7,060	810	2,790		310	980	102	280	96	250	520	120		95	1,010	110
12-Oct	7,060	470	750		370	980	108	630	96	220	500	120		95	980	105
13-Oct	7,060	1,100	1,310		370	1,060	105	630	96			90	980	95	850	100
14-Oct	7,060	480	760		410		74	730	102			100	1,110	105		100
15-Oct	7,060	850	1,120		370		75	680	102			100	1,110	108	850	105
16-Oct	6,750	520	780		260		72	730	105			100	1,110	108	850	105
17-Oct	7,060	1,100	2,490		310		80	780	105	170	310	100	1,110	108	850	105
18-Oct	7,060	220	490		310		72	680	105	150	220	100	1,110	108		100
19-Oct	7,060	80	300			980	68		105	170	300	100	1,110	108	910	105
20-Oct	5,840	220	1,060			980	62		90	150	260	90	980	88	780	100
21-Oct	4,840	430	1,380			980	60		85	130	260	80		85	850	100
22-Oct	3,170	420	1,080			320	52		72	20	60	60		82	660	100
23-Oct	5,260	1,240	1,990		80	370	60		84		190	90		100	1,250	115
24-Oct	4,840	1,190	1,910		150	370	58		82		190	90		100	1,250	115
25-Oct	2,710	980	1,390		150	520	100		48			90		58	390	52
26-Oct	2,710	540	920		150		92	200	90			50		58	390	52
27-Oct	2,380	430	670		100		70	160	75			50		55	300	52
28-Oct	2,710	280	980		150		90	240	85			50		67	350	52
29-Oct	2,710	590	1,010		150		85	330	82			50		67	350	52
30-Oct	2,940	450	670		150	430	95	80	78			50		68	350	52

Role of sediment in the design and management of irrigation canals

Date	Water inflow	Sediment inflow		Water outflow and set points													
		> 63 µm	Total	SS9A	SS9C	SS9B	WLR-1	SS9D	WLR-2	S9-T2	S9-T1	WLR-3	SS9E	WLR-4	SS9F	WLR-5
	l/s	ppm	ppm	l/s	l/s	l/s	cm	l/s	cm	l/s	l/s	cm	l/s	cm	l/s	cm
31-Oct	2,380	270	480			130	60		60			50		68	350	52
1-Nov	2,490	1,430	2,600			130	62		60			50		67	350	52

Blank = no flows, WLR = Water Level Regulator

Appendix C: Predictor-corrector method

Steps used in the model SETRIC to compute gradually varied flow profile using predictor corrector method.

- compute the derivative (dh/dx) at point $x = x_i$ (S_0, S_{f1}, and $Fr_1{}^2$ are known):

$$\left(\frac{dh}{dx}\right)_i = \frac{(S_0 - S_f)_i}{(1 - F_r{}^2)_i} \tag{C.1}$$

- calculate the water depth h_{i+1} at point $x = x_{i+1}$ with $(dh/dx)_i$:

$$\left(h_{i+1}\right)_1 = h_i + \left(\frac{dh}{dx}\right)_i (x_{i+1} - x_i) \tag{C.2}$$

- with the new water depth value (h_{i+1}) calculate $\left(S_f\right)_{i+1}$ and Fr_{i+1};
- calculate the derivative at point $x = x_{i+1}$:

$$\left(\frac{dh}{dx}\right)_{i+1} = \frac{(S_0 - S_f)_{i+1}}{(1 - F_r{}^2)_{i+1}} \tag{C.3}$$

- calculate the mean derivative:

$$\left(\frac{dh}{dx}\right)_{mean} = \frac{\left(\dfrac{dh}{dx}\right)_i + \left(\dfrac{dh}{dx}\right)_{i+1}}{2} \tag{C.4}$$

- calculate the new value of h_{i+1} by:

$$\left(h_{i+1}\right)_2 = h_i + \left(\frac{dh}{dx}\right)_{mean} (x_{x+1} - x_i) \tag{C.5}$$

- check the accuracy of predicted value (e = degree of accuracy desired):

$$\left| \left(h_{i+1}\right)_1 - \left(h_{i+1}\right)_2 \right| \le e \tag{C.6}$$

The procedure given above can be repeated if necessary, and may also be used for short sections of canals with changing width and shape (non-prismatic canals). The size of length step depends upon the relative change in water depth and desired accuracy.

Appendix D: Galappatti's depth integrated model for non-equilibrium sediment transport

Galappatti (1983) has developed a depth integrated suspended sediment transport model based on an asymptotic solution for the two-dimensional convection equation in the vertical plane. Among the depth integrated model for suspended sediment transport this model has two advantages over others; firstly no empirical relation has been used during the derivation of the model and secondly all possible bed boundary conditions can be used (Z. B. Wang, *et al.*, 1986). Moreover it includes the boundary condition near the bed, hence an empirical relation for deposition/pick-up rate near the bed is not necessary (Ribberink, 1986). The theoretical background, assumptions and limitations of the model is briefly given in the subsequent paragraphs.

The partial differential equation that governs the transport of suspended sediment by convection and turbulent diffusion under gravity is given by (Galappatti, 1983):

$$\frac{\partial c}{\partial t}+u\frac{\partial c}{\partial x}+v\frac{\partial c}{\partial y}+w\frac{\partial c}{\partial z}=w_s\frac{\partial c}{\partial z}+\frac{\partial}{\partial x}(\varepsilon_x\frac{\partial c}{\partial x})+\frac{\partial}{\partial y}(\varepsilon_y\frac{\partial c}{\partial y})+\frac{\partial}{\partial z}(\varepsilon_z\frac{\partial c}{\partial z}) \quad (D.1)$$

Neglecting the diffusion terms other than vertical, equation (D.1) for a two dimensional flow in the vertical plane becomes:

$$\frac{\partial c}{\partial t}+u\frac{\partial c}{\partial x}+w\frac{\partial c}{\partial z}=w_s\frac{\partial c}{\partial z}+\frac{\partial}{\partial z}(\varepsilon_z\frac{\partial c}{\partial z}) \quad (D.2)$$

Galappatti (1983) assumed a flow field as shown in Figure D.1 for the derivation of his model for non-equilibrium sediment transport. The definition of the symbols used in the figure are:

h	= flow depth above reference level (m)
h_0	= total flow depth (m)
x, z	= length coordinates (m)
u, w	= velocity component in x, and z direction (m/s)
z_s	= water surface elevation (m)
Δz	= $h_0 - h$ (m)
z_a	= bed elevation (m)

The boundary conditions applied for the depth-integrated model are as follows:
- the concentration at the upstream boundary at each time step is known;
- no sediment flux across the surface, i.e. $(w_s c+\varepsilon_z\frac{\partial c}{\partial z})_{surface}=0$;

the bed-load concentration C(bed) as function of flow and sediment parameters.

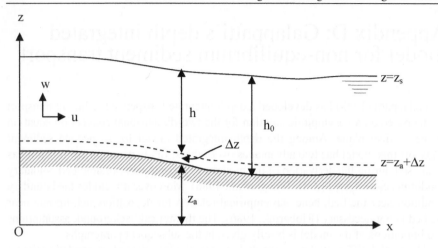

Figure D.1   Flow field (Galappatti, 1983).

The boundary condition for the bed is not applied at the bed ($z = z_a$), but at a small distance Δz from the bed $z = z_a + \Delta z$. There are several types of bed boundary conditions that could be applied. In Galappatti's analysis only one type, i.e. the value of concentration near the bed $C_{(bed)}$ at the level $z = z_a + \Delta z$ is applied. It has been assumed that $C_{(bed)}$ at $z = z_a + \Delta z$ is known in terms of local flow and sediment parameters. In other words, $C_{(bed)}$ is known in advance. Moreover, the concentration at any section is expressed as the depth averaged concentration as shown in Figure D.2. Equation (D.2), if integrated vertically with the preset boundary conditions gives the depth averaged equation.

$$\frac{\partial}{\partial t}(h\overline{c}) + \frac{\partial}{\partial x}(h.\overline{u}\overline{c}) = E \tag{D.3}$$

$$h.\overline{u}\overline{c} = \int_{Z_A + \Delta Z}^{Z_A + \Delta Z + h} ucdz \tag{D.4}$$

where

$\overline{c}$ = mean concentration (m^3 sediment / m^3 water)
E = entrainment rate
h = depth of flow (m)
$\overline{u}$ = mean flow velocity (m/s)

For the asymptotic solution of the convection-diffusion equation the following two assumptions are made:

$$\frac{UH}{Lw_s} = \delta \ll 1$$

$$\frac{H}{w_s T} = \delta \ll 1 \tag{D.5}$$

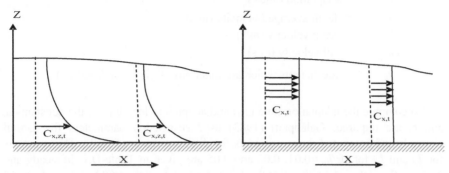

Figure D.2   Schematization of 2-D and depth-integrated suspended sediment transport models.

This implies that the time taken for a particle to settle (t = H/w$_s$) is much smaller than the time it takes to be convected along a distance L (T = L/U). Now, the general solution for the sediment concentration in non-equilibrium conditions in terms of depth averaged variables is given by:

$$\gamma_{11}\bar{c}_e = \gamma_{11}\bar{c} + \gamma_{21}\frac{h}{w_s}\frac{\partial \bar{c}}{\partial t} + \gamma_{22}\frac{uh}{w_s}\frac{\partial \bar{c}}{\partial x} \tag{D.6}$$

$$\bar{c}_e = \bar{c} + T_A\frac{\partial \bar{c}}{\partial t} + L_A\frac{\partial \bar{c}}{\partial x} \tag{D.7}$$

with

$$T_A = \frac{\gamma_{21}}{\gamma_{11}}\frac{h}{w_s} = \frac{w_s}{u^*}\frac{h}{w_s}\exp(f) \tag{D.8}$$

$$\frac{T_A u^*}{h} = \exp(f) \tag{D.9}$$

$$L_A = \frac{\gamma_{22}}{\gamma_{11}}\frac{\bar{u}h}{w_s} = \frac{w_s}{u_*}\frac{\bar{u}h}{w_s}\exp(f) \tag{D.10}$$

$$L_A = \frac{Vh}{u^*}\exp(f) \tag{D.11}$$

$$f = \sum_{i=1}^{4}(a_i + b_i\frac{u_*}{V})(\frac{w_s}{u_*})^{i-1} \tag{D.12}$$

where

a_i, b_i	= constants
$\bar{c}_e$	= concentration of suspended load in equilibrium condition (ppm)
$\bar{c}$	= concentration of suspended load at distance x (ppm)
L_A	= adaptation length (m)

T_A = adaptation time (s)

$\bar{u}$ = depth averaged velocity (m/s)

V = mean velocity (m/s)

w_s = fall velocity (m/s)

γ_{ij} = coefficients which are functions of ($w_s, u_*, \bar{u}, h$ and z_a)

To compute the adaptation time (T_A) and adaptation length (L_A), the constants a_i and b_i are required. Galappatti (1983) used suspension parameters for natural channels and computed those constants for different ratios of $\Delta z/h_0$. The constants for T_A and L_A for $\Delta z/h_0$ =0.01, 0.02 and 0.05 are given in Table D.1. In steady and uniform flow, the adaptation length (L_A) and adaptation time (T_A) are constant and have straight characteristics in the x-t plane. In gradually varied flow the water depth and velocity changes in each length step. Therefore L_A and T_A are not constant along the canal. Furthermore, the magnitude of $|c-c_e|$ decreases exponentially with L_A and T_A. They are defined as the interval (both in length and time) required for the actual concentration to approach the mean concentration. The adaptation length and adaptation time represent the length scale and time scale respectively.

Table D.1   Value of a and b for different $\Delta z/h_0$ (Galappatti, 1983).

		A_1	b_1	a_2	b_2	a_3	b_3	a_4	b_4
$\Delta z/h_0 =$	T_A	1.978	0.000	-6.321	0.000	3.256	0.000	0.193	0.000
0.01	L_A	1.978	0.543	-6.325	-3.331	3.272	0.400	0.181	1.790
$\Delta z/h_0 =$	T_A	1.788	0.000	-5.779	0.000	2.860	0.000	0.226	0.000
0.02	L_A	1.789	0.570	-5.783	-3.000	2.872	0.560	0.217	1.430
$\Delta z/h_0 =$	T_A	1.486	0.000	-4.999	0.000	2.306	0.000	0.247	0.000
0.05	L_A	1.486	0.576	-5.002	-2.416	2.314	0.720	0.242	0.910

For steady sediment flow, equation (D.7) becomes:

$$\bar{c}_e = \bar{c} + L_A \frac{\partial \bar{c}}{\partial x} \tag{D.13}$$

Integration results:

$$\int_0^x dx = L_A \int_{c_0}^{\bar{c}} \frac{d\bar{c}}{\bar{c}_e - \bar{c}} \tag{D.14}$$

$$\bar{c} = \bar{c}_e - (\bar{c}_e - \bar{c}_0) \exp^{-\frac{x}{L_A}} \tag{D.15}$$

Symbols

Symbol	Meaning	Unit
A	Cross sectional area	(m)
B, b	Channel bottom width	(m)
B_s	Water surface width	(m)
C, c	Sediment concentration	(m)
C_e. c_e	Equilibrium concentration	(ppm by mass)
C_0, c_0	Concentration at length x = 0	(ppm by mass)
C	Chezy's coefficient	$(m^{1/2}/s)$
C_d	Discharge coefficient	-
d_{50}	Median particle size	(m)
D	Diameter	(m)
D_*	Dimensionless particle diameter	-
E	Total energy	(Nm/N)
F_g	Grain Froude number	-
F_{gr}	Dimensionless mobility parameter	-
F_{gcr}	Critical grain Froude number	-
F_r	Froude number	-
G	Acceleration due to gravity	-
G_{gr}	Dimensionless mobility parameter	-
h, y	water depth	(m)
K_s	Nikuradse's roughness height	(m)
K'_s	Nikuradse's grain roughness height	(m)
K''_s	Nikuradse's bed form roughness height	(m)
K_{se}	Effective equivalent roughness	(m)
l, L	Length	(m)
L_A	Adaptation length	(m)
M	Side slope (1V:mH)	-
N	Manning's coefficient	$s/m^{1/3}$
P	Porosity	-
P	Wetted perimeter	(m)
Q	Flow discharge	(m^3/s)
Q_s	Sediment discharge per unit width	(m^2/s)
q_s	Total sediment discharge	(m^3/s)
R	Hydraulic radius	(m)
Re_*	Particle Reynolds number	-
R_g	Grain Reynolds number	-
s	Relative density of sediment	-
S_f	Friction slope	-
S_0	Bed slope	-
t, T	Time	(s)
T	Excess bed shear stress parameter	-
T_A	Adaptation time	(s)
u,v,w	Component of velocity in X, Y, Z directions	(m/s)

Symbol	Meaning	Unit
U_*	Shear velocity	(m/s)
U_{*cr}	Critical shear velocity	(m/s)
V, U	Mean velocity	(m/s)
ω_s	Fall velocity	(m/s)
y_c	Critical depth	(m)
y_n	Normal depth	(m)
α	Correction factor	-
γ	Specific weight	(N/m³)
γ_d	Form factor	-
γ_r	Ripple presence	-
Δ	Bed form height	(m)
λ	Bed form length	(m)
ε	Mixing coefficient	(m²/s)
σ_s	Geometric standard deviation	-
τ	Shear stress	(N/m²)
τ_{cr}	Critical shear stress	(N/m²)
μ	Dynamic viscosity	(kgm/s)
ν	Kinematic viscosity	(m²/s)

Glossary

Adequacy	The ratio of the amount delivered to the amount required.
Adjustable flow	Irrigation flow regulated at a specific discharge (flow rate).
Aggradation	Build-up or raising of the canal bed due to sediment deposition.
Alignment	The course along which the centre line of a canal or drain is located.
Alluvial canal	A canal wholly in unconsolidated material deposited by stream and whose processes are controlled by the flow and boundary interactions.
Alluvial soil	A soil formed from deposits by rivers and streams (alluvium).
Analytical model	System of mathematical equations that are the algebraic solutions of the fundamental equations.
Backwater	The increase in water surface elevation relative to the elevation occurring under normal canal conditions, induced due to control structures or constrictions at the downstream end.
Bank protection	The process by which the bank is protected from erosion by lining or by retarding the velocity along the bank; device to reduce scour by flowing water.
Bed erosion	Deepening of a canal by the gradual wearing away of bed material, mainly due to the forces of flowing water.
Bed forms	A recognisable relief feature on the bed of the canal such as ripple, dune or bar.
Bed material	Material with particle sizes that are found in significant quantities in that part of the bed affected by transport.
Bed material load	That part of total loads that is composed of particle sizes present in appreciable quantities in the shifting part of the streambed.
Bed or bottom slope	Difference in bed elevation per unit horizontal distance in the flow direction.
Bed profile	Shape of the bed in a vertical plane; longitudinally or transversely.
Bed shear stress	The way in which currents transfer energy to the channel bed.
Bed load	Material moving on or near the streambed by rolling, sliding and skipping.
Bifurcation	Location where a river or canal separates in two or more reaches, sections or branches.
Bottom width	Dimension of a canal bottom measured normal to the flow direction.
Boundary conditions	Physical conditions (hydraulic and/or others) used as boundary input to \physical or numerical models.
Boundary layer	The flow region next to a solid boundary where the flow field is affected by the presence of the boundary and where friction plays an essential part.

Canal	A man-made or natural channel used to convey water.
Chézy coefficient	Resistance coefficient for open channel flows; coefficient is a function of the relative roughness and Reynolds number.
Cohesion	The mutual attraction of soil particles due to strongly attractive forces.
Cohesionless soil	Soil that has a little tendency to stick together whether wet or dry, such as sands and gravel.
Cohesive sediment	Sediment material of very small sizes (less than 50 μm) for which cohesive bonds between particles are significant and affect the material properties.
Concentration	The ratio of mass of dry sediment in a water-sediment mixture to the mass of the mixture and reported in milligram per meter. This can be converted to PPM by multiplying 1000 if the PPM is up to 15000.
Continuity	The fundamental law of hydrodynamics, which states that for incompressible flows independent of time, the sum of the differential changes in flow velocities in all directions must be zero.
Control	Physical properties of a canal, which determine the relationship between stage and discharge at a location in that canal.
Control section	A section in an open canal where critical flow conditions take place resulting in a distinct relationship between water level and discharge; the concept of 'control' and 'control section' are used with the same meaning.
Critical depth	Depth in a canal of specified dimensions at which the mean specific energy is a minimum for a given discharge; the flow is critical.
Critical flow	Flow for which the specific energy is minimum for a given discharge; Froude number will be equal to unity and surface disturbances will not travel upstream.
Cross-section	Section of a stream normal to the flow direction bounded by the wetted perimeter and the free surface.
Datum	Any permanent plane or surface used as a reference datum to which elevations are referred.
Degradation	Lowering of canal bed elevation resulting from the erosion of sediments; downward and lateral erosion.
Density	Mass (in kg) per unit of volume.
Design discharge	A specific value of the flow rate that after the frequency and the duration of exceedance have been considered, is selected for designing the dimensions of a structure or a system or a part thereof.
Design flow	Flow on which the cross section of a canal is determined.
Desilting basin	Canal section with very low velocity, forming deposition of the sediment carried by the water.
Diameter intermediate axis	The diameter of a sediment particle determined by direct measurement of the axis normal to a plane representing the longest and shortest axis.

Diameter, fall	The diameter of a sphere with a specific density of 2.65 and the same standard fall-velocity as the particle. (Also sometimes known as sedimentation diameter).
Diameter, nominal	The diameter of a sphere of the same volume as the given particle (Venoni 1975).
Diameter, sieve	The size of the sieve opening through which the particle of sediment will just pass.
Diffusion	The process whereby particles intermingle as the result of spontaneous movement and move from a region of higher concentration to one of lower concentration.
Discharge	Volume of water per unit time flowing along a canal.
Discharge regulator	Structure regulating the flow from one canal to another.
Diversion works	Diversion works are composed by the structures and hydro-mechanical equipment designed to divert and control the required discharge from a given river. Generally, the diversion works include the intake, a diversion dam and sluiceway, a spillway, an approach canal, training works (dikes, levees, protection, etc.)
Drawdown curve	Water surface profile, when water surface slope is larger than bottom slope.
Drop	A rapid change of bed elevation, also called a step.
Dunes	Types of bed form indicating significant sediment transport over a sandy bed.
Dynamic equilibrium	Short-term morphological changes that do not affect the morphology over a long period.
Dynamic viscosity	The ratio between the shear stress acting along any plane between neighbouring fluid elements and the rate of deformation of the velocity gradient perpendicular to this plane
Eddy viscosity	Name for the momentum exchange coefficient; also called 'eddy coefficient'.
Energy or head loss	Difference in total energy between two cross-sections.
Equity	Criterion of the share for each individual or group that is considered fair by all system members; the measure can be defined as the delivery of a fair share of water to users throughout a system; the spatial uniformity of the ratio of the delivered amount of water to the required (or scheduled) amount
Erosion	The process by which soil is washed or otherwise moved by natural factors from one place to another; the gradual wearing away of canal beds by flowing water.
Flow control method	In general, a regulation method for irrigation structures to maintain a specific flow condition in the irrigation system.
Free flow	Flow through a canal, which is not affected by the level of the downstream water.

Friction	Boundary shear resistance of the wetted surface of a canal, which opposes the flow of water; process by which energy is lost through shear stress.
Friction coefficient	Coefficient used to calculate the energy gradient by friction.
Geometric standard deviation	Geometric standard deviation of bed particle size also known as gradation is given by .
Gradually varied flow	Flow characterized by relatively small changes in velocity and pressure distributions over a short distance.
Head-discharge relationship	Curve or table which gives the relation between the head and the discharge in an open channel at a given cross-section for a given flow condition, e.g. steady, rising or falling.
Headworks	Headworks are composed by the structures and equipment located at the upstream end of a water supply system, designed to direct the diverted discharge from the river into conveyance system.
Hydraulic mean depth	Depth obtained by dividing the cross-sectional area by the free surface or top width.
Intake	Intake means the structures and hydro-mechanical equipment designed to abstract water from the river and make up the essential component of any river diversion.
Idle length	Length of a canal from the intake to the first off-take.
Karman constant	'Universal' constant of proportionality between the Prandtl mixing length and the distance from the boundary. Experimental results indicate that $\kappa = 0.40$.
Kinematic viscosity	Dynamic viscosity divided by the fluid density.
Laminar flow	Flow characterized by fluid particles moving along smooth paths in thin layers (laminas), with one layer gliding smoothly over an adjacent layer and not influenced by adjacent layers perpendicular to the flow direction.
Longitudinal section	Vertical section along the centre line of a canal, it shows the original and the final levels.
Main canal	Irrigation canal taking water from the source and conveying the water to the secondary canals.
Maintenance	Operations performed in preserving irrigation or drainage canals and drain pipes, hydraulic structures, service roads and works in good or near original conditions. Repairs are part of maintenance; the regular, continuous inspection and repair of irrigation and drainage systems.
Mathematical model	Model that simulates a system's behaviour by a set of equations, perhaps together with logical statements, by expressing relationships between variables and parameters.
Mean depth	Average depth of a canal, being the cross sectional area divided by the surface or top width.
Mean velocity	Average velocity in a canal, being the discharge divided by the cross sectional (wetted) area.

Modelling	Simulation of some physical phenomenon or system with another system believed to obey the same physical laws or rules in order to predict the behaviour of the former by experimenting with the latter.
Modular limit	Defined as this limiting submergence ratio for a particular flow module, which causes no more than a 1percent deviation in the upstream head reading for a given discharge
Momentum exchange coefficient	Apparent kinematic (eddy) viscosity in turbulent flows; analogous to the kinematic viscosity in laminar flows. Momentum exchange coefficient is proportional to the shear stress divided by the strain rate.
Non-uniform flow	Flow that varies in depth, cross sectional area, velocity and hydraulic slope from section to section.
Normal depth	Uniform equilibrium open channel flow depth; depth at a given point in a canal corresponding to uniform flow, water surface and bed are parallel.
Numerical modelling	Refers to the analysis of physical processes (e.g. hydraulic) using computational models.
Off-take	Structure with or without gates, that conveys water to a secondary canal or tertiary unit.
One-dimensional flow	Neglects the variations and changes in velocity and pressure transverse to the main flow direction.
One-dimensional model	Model defined with one spatial coordinate, the variables being averaged in the other two directions.
Open canal	Natural or man-made structure that contains, restricts and directs the flow of water. The surface of the water is open to the atmosphere, and therefore, the flow is referred to as free flow. The design of canals includes the solution of relationships between bed and bank roughness, canal geometry, and flow velocity. Free surface flows are driven by gravity and they can vary in both time and space.
Overflow structure	Structure with water flowing over its crest.
Particle size distribution	The fractions of clay, silt and sand particles in a soil.
ppm	Abbreviation for parts per million.
Quasi-steady flow	Quasi-steady flow is a condition in which the discharge can be considered to be constant throughout the length (x) at a particular time interval, although it may be changing with time (t).
Rating curve	Graphic or tabular presentation of the discharge or flow through a structure or canal section as a function of water stage (depth of flow).
Regulator	Structure to set (regulate) water levels and/or discharges in an irrigation network
Response time	Time lag, i.e., the time needed for a canal network to reach a new steady state after a change in water level or discharge.

Roughness coefficient	Factor in formulae for computing the average flow velocity in open channels that represents the effect of roughness and other geometric characteristics of the channel upon the energy losses; e.g. the de Chézy, Manning or Strickler coefficients.
Sand	Sediment particles, mainly quartz, with a diameter of between 0.062 mm and 2 mm, generally classified as fine, medium, coarse or very coarse.
Sand trap	Enlargement in a canal where the velocity drops so that any sand that it carries can settle and be removed.
Scour	Removal of bed material by the eroding power of a flow of water; erosive action, particularly, pronounced local erosion by fast flowing water that excavates and carries away material from the bed and banks.
Sediment	Particle derived from rocks, biological materials, of chemical precipitants that are transported by, suspended in or deposited by flowing water.
Sediment concentration	The ratio of the mass (or volume) of the dry sediment in a water/sediment mixture to the total mass (or volume) of the suspension.
Sediment discharge	The mass or volume of sediment passing a stream cross-section per unit time.
Sediment load	A general term that refers to material in suspension and/or in transport. It is not synonymous with either discharge or concentration.
Sediment transport	Movement of sediment transported in any way by a flow; from the point of transport it is the sum of suspended and bed-load transported; from the point of origin it is the sum of bed material load and the wash load.
Sediment transport capacity	Ability of a stream to carry a certain volume of sediment per unit time for given flow conditions, also called the sediment transport potential.
Sediment yield	Total sediment outflow including bed-load and suspension.
Sedimentation	Deposition of sediments in canals due to a decrease in velocity and corresponding reduction in the size and amount of sediment that can be carried.
Set point	The target value or desired output.
Settling basin	Small basin placed at selected points along a conduit, open canal or subsurface drain to collect sand and silt, and also to afford an opportunity for inspection of the operation. See silt basin.
Side slope	Slope of the side of a canal with the horizontal; tangent of the angle with the horizontal; the ratio of the horizontal and vertical components of the slope.
Silt	Sediment particles with a grain size between 0.004 mm and 0.062 mm, i.e. coarser than clay particles but finer than sand.
Silt/sediment pocket	Small basin placed at selected points along a canal to collect sand and silt.

Silt clearance	Removal of silt deposited in a canal section above the design bed levels; the general term also includes bank trimming in the case of constriction of width by silting.
Silting	Process of accretion or rising of the canal bed by depositing of sediment in the flow. Also called 'accretion of silt. Building of silt layers on canal sides is referred to as silting, but not as accretion.
Simulation	Representation of a physical system by a computer or a model that imitates the behaviour of the system; a simplified version of a situation in the real world.
Stable canal	Canal in which the bed and sides remain stable over a significant period of time and in which scour and deposition is minimal.
Steady flow	Flow in which the depth and velocity remain constant with respect to time.
Steady state	Fluid motion in which the velocity at every point of the field is independent of time in either magnitude or direction; flow condition in which the input energy equals the output energy.
Sub-critical flow	The flow in an open canal is sub-critical if the flow depth is larger than the critical depth. A flow for which the Froude number is less than unity; surface disturbances can travel upstream.
Suspended concentration	The in time-average ratio of the mass (or volume) of the dry sediment in a water/sediment mixture to the total mass (or volume) of the mixture; also average of the mean (time average) suspended concentration over the entire area.
Suspended load	Transported sediment material maintained into suspension by turbulence of the flow for considerable periods and without contact with the bed; the velocity of the load is almost the same as that of the flow; it is part of the total sediment transport.
Suspended sediment	Sediment that is carried in suspension by the turbulent component of the liquid or by Brownian movement.
Tertiary off-take	A discharge regulator on the secondary or primary canal to supply a tertiary unit.
Time lag	Time needed for a canal network to reach a new steady state after a change in water level or discharge.
Top width	Width of the canal measured at the water surface.
Total energy head	Sum of the elevation of the free water surface above a horizontal datum in a section and the velocity head at that section.
Total load (origin)	Total load comprises the 'bed material load' (including suspended load) and the 'wash load'.
Total load (transport)	The total load consists of the 'bed-load' and 'suspended load' (including wash load).
Tractive force	The force exerted by flowing water on the bed and banks of a canal and tangential to the flow direction.
Turbulent flow	Fluid particles move in very irregular paths, causing exchange of momentum from one part of the flow to another.

Uniform flow	Flow with no change in depth or any other flow characteristic (wetted area, velocity or hydraulic gradient) along a canal.
Unsteady flow	Flow in which the velocity changes, with time, in magnitude or direction.
Validation	The comparison between model results and prototype data, to validate a numerical model. Validation is carried out with prototype data that are different from those used for calibration and verification of the model.
Velocity	The rate of movement at a certain point in a specified direction.
Velocity head	The energy per unit weight of water in view of its flow velocity; square of the mean velocity divided by twice the acceleration due to gravity.
Volume Control	Flow control method with the set point in the middle of a canal reach.
Wash load	Portion of the suspended load with particle sizes smaller than those found in the bed; in near-permanent suspension and transported without deposition; the amount of wash load transported through a section is independent of the transport capacity of the flow.
Water-level	Elevation of the free water surface relative to a datum.
Water-level regulator	Structure to regulate the water level also known as cross regulator
Water-surface slope	Difference in elevation of the water surface per unit horizontal distance in the flow direction.
Wetted perimeter	The length of wetted contact between water and the solid boundaries of a cross-section of an open channel; usually measured in a plane normal to the flow direction.

Samenvatting

Sediment transport in irrigatie kanalen

Sediment transport is een belangrijke factor in de verdere ontwikkeling van irrigatie omdat het in belangrijke mate de duurzaamheid van een irrigatie systeem bepaalt, speciaal als er sprake is van onbeklede kanalen in alluviale gronden. Onderzoeken naar het gedrag van sediment begonnen in 1895 toen Kennedy zijn theorie over bodemvormende afvoeren publiceerde. Later zijn nog andere theorieën ontwikkeld die over de hele wereld worden gebruikt. Alle theorieën veronderstellen eenparige en permanente stroming en proberen om de afmetingen van een kanaal te vinden die voor een bepaalde afvoer en sediment transport stabiel zijn. In het verleden werden irrigatie systemen ontworpen voor de bescherming van de landbouw, maar met weinig debietregeling, vandaar dat de permanente en eenparige stromingsvoorwaarden in zekere mate gerealiseerd zijn.

De moderne irrigatie systemen zijn meer en meer op de vraag naar water gebaseerd wat betekent dat de watertoevoer in een kanaal door de behoefte van de gewassen wordt bepaald. Dienovereenkomstig is de aanvoer in het kanaalnetwerk niet constant, aangezien de behoefte van de gewassen met het klimaat en de groeistadia verandert. Ook is de aanvoer van sediment in de meeste systemen niet constant tijdens het irrigatie seizoen. De situatie is nog slechter voor systemen die direct van een rivier aftappen en waar de schommelingen in de rivierafvoer een direct effect op de aanvoer van water en sediment hebben.

De conventionele ontwerp methoden kunnen het sediment transport in een kanaal niet nauwkeurig voorspellen, ten eerste door de niet-permanente en niet-eenparige waterstroming en ten tweede door de veranderingen in de sediment aanvoer. Vandaar dat het werkelijke gedrag van een kanaal sterk afwijkt van de aannamen die tijdens het ontwerp gemaakt zijn en in veel gevallen moeten zeer hoge onderhoudskosten worden opgebracht om de sediment problemen aan te pakken.

Een irrigatie systeem zou niet alleen de juiste hoeveelheid water, op het correcte moment en op het gewenste niveau aan de gewassen moeten kunnen leveren, maar ook zou het minstens zijn beheer- en onderhoudskosten moeten terugverdienen. De terugbetaling van de kosten is in zekere mate gekoppeld aan het niveau van dienstverlening van de irrigatie organisatie en de uitgaven voor onderhoud van het systeem. Ervaringen uit het verleden in Nepal hebben aangetoond dat modernisering van bestaande irrigatie systemen om het niveau van dienstverlening te verbeteren ook heeft geleid tot verhoging van de beheer- en onderhoudskosten. Deze kosten zijn, in sommige gevallen hoog vergeleken met de over het algemeen beperkte mogelijkheden van de watergebruikers en landbouwers om deze kosten terug te betalen. Het streven om de systemen eerlijker, betrouwbaarder en flexibeler te maken heeft geleid tot de introductie van nieuwe waterbeheer- en wateraanvoer systemen die als ze niet zorgvuldig ontworpen worden het sediment transport ongunstig beïnvloeden. In veel systemen heeft ongewenste sedimentatie en/of erosie in kanalen niet alleen de beheer- en onderhoudskosten laten stijgen maar ook de betrouwbaarheid van de geleverde diensten doen afnemen.

Irrigatie ontwikkeling in Nepal en het studiegebied

Nepal ligt in Zuid-Azië tussen China en India en is geheel door land omgeven. Het land ligt tussen 26°22' N en 30°27' N breedte en 80°4' E en 88°12' E lengte. De vorm is ruwweg rechthoekig en de oppervlakte is 147181 km^2. Nepal is 885 km lang, maar de breedte neemt af naar het Westen. De gemiddelde Noord-Zuid breedte is 193 km. Nepal is een uitgesproken bergachtig land met hoogten van 64 m+GZN (Boven Gemiddeld Zee Niveau) in Kechana, Jhapa tot 8.848 m+GZN bij de Mount Everest, de hoogste berg ter wereld, en dat binnen een spanwijdte van 200 km. Nepal heeft een landbouwgebied van 2,64 miljoen ha, waarvan tweederde (1,8 miljoen ha) potentieel geïrrigeerd kan worden. Momenteel heeft 42% van het landbouwgebied één of andere vorm van irrigatie, maar slechts 41% daarvan ontvangt het hele jaar irrigatie water. De bestaande irrigatie systemen dragen voor ongeveer 65% bij aan de huidige landbouwproductie van het land.

Nepal heeft een lange geschiedenis van geïrrigeerde landbouw. De meeste grootschalige irrigatie systemen liggen in de zuidelijke alluviale vlakte (Terai). De kanalen zijn onbekleed en het meegevoerde sediment vormt een integraal onderdeel van het aangevoerde irrigatie water. De systemen zijn vooral aanvoer georiënteerd en geven een kleine watergift voor intensieve landbouw. Gezien de toenemende concurrentie tussen de verschillende water sectoren en de geringe prestaties van de irrigatie systemen worden vele gemoderniseerd. Het Sunsari Morang Irrigatie Systeem is één van de systemen dat onder de modernisering valt. Dit gebied is als studie object voor dit onderzoek gebruikt. Een verbeterd inzicht in het sediment transport proces, onder veranderende stroming en sediment omstandigheden, een zich wijzigende beheeromgeving en verschillende onderhoud schema's, zal erg nuttig zijn om de systemen uit de huidige vicieuze cyclus van bouw – aftakeling – rehabilitatie te halen.

Het Sunsari Morang Irrigatie Systeem ligt in de oostelijke Terai. De Koshi rivier is de bron van irrigatie water. Een zijdelingse inlaat, een ruim 50 km lang hoofdkanaal met een capaciteit van 45,3 m^3/s voor de wateraanvoer en 10 secundaire kanalen en lagerorde kanalen met verschillende capaciteiten voor de waterdistributie zijn aangelegd om een gebied van 68.000 ha te irrigeren. Het systeem werd voor het eerst in 1975 gebruikt, maar had al vrij snel ernstige problemen met de wateraanvoer en sedimentatie in het kanaalnetwerk. Vandaar dat vanaf 1978, slechts 3 jaar na de ingebruiknname de rehabilitatie- en modernisering werkzaamheden begonnen. Tijdens de modernisering is het hoofdinlaatwerk verplaatst om de aftap van water te verbeteren en de sediment aanvoer te verminderen. Bovendien is een zandvang met baggermachines voor het continue verwijderen van het sediment aan het begin van het hoofdkanaal aangelegd. Behalve dat zijn de verdere ontwikkeling van het gebied en de modernisering van het bestaande kanaal netwerk gaande en na de derde en laatste moderniseringsfase (1997 – 2002) waren ongeveer 41.000 ha ontwikkeld.

Sediment transport onderzoek

Het doel van dit onderzoek is het onderkennen van de relevante aspecten van sediment transport in irrigatie kanalen en het formuleren van een ontwerp methode en beheersysteem voor irrigatie systemen in Nepal, gezien het sediment transport. Het onderzoek heeft zich vooral gericht op de huidige ontwerp methoden van irrigatie systemen in Nepal en hun doeltreffendheid ten aanzien van het sediment transport. Het effect van het beheer en onderhoud op het sediment transport is geanalyseerd op basis van gegevens van het SMIS (Sunsari Morang Irrigatie Systeem). Vervolgens is een verbeterde ontwerp methode in verband met het sediment transport in irrigatie kanalen geformuleerd. Het computer model SETRIC is gebruikt om de relatie tussen sediment transport, het ontwerp en het beheer te bestuderen en de voorgestelde ontwerp methode voor irrigatie kanalen te evalueren op basis van gegevens van het irrigatie systeem.

De wiskundige formulering van het sediment transport in een irrigatie kanaal is gebaseerd op eerder uitgevoerd onderzoek, in het bijzonder het onderzoek van Méndez en de totstandkoming van het computer model SETRIC. Vervolg onderzoeken, analyses, verbeteringen en verificatie door Paudel, Ghimire, Orellana V., Via Giglio en Sherpa zijn ook gebruikt. Het model SETRIC is geverifieerd en waar nodig verbeterd en is daarna gebruikt om het irrigatie systeem te analyseren en om verbeteringen in het ontwerp en beheer vanuit het standpunt van sediment transport voor te stellen.

Evaluatie van de ontwerpfactoren

De factoren die de ruwheid van een irrigatie kanaal beïnvloeden zijn bestudeerd en een voorstel voor een meer rationele bepaling van de ruwheid is uitgewerkt op basis van de beschikbare kennis. De ruwheid van de oevers is afhankelijk van de vorm en grootte van het materiaal, de begroeiing en onregelmatigheden van het oppervlak, terwijl de ruwheid van de bodem een functie is van de vorm en grootte van het bodem materiaal en onregelmatigheden van het oppervlak (bodem vormen in het geval van alluviale kanalen). Voor de bepaling van de bodem ruwheid zijn meestal twee methoden in gebruik, namelijk een methode die gebaseerd is op hydraulische grootheden (waterdiepte, stroomsnelheid en bodem materiaal) en de methode die gebruik maakt van bodem vorm en korrel gerelateerde parameters. Dit onderzoek maakte gebruik van de methode die gebaseerd is op de bodem vorm en korrel gerelateerde parameters zoals door van Rijn is voorgesteld. Op dezelfde manier is voor de bepaling van de ruwheid van de oevers de invloed van onregelmatigheden van het oppervlak en begroeiing uitgedrukt door de onderhoud toestand te verdelen in ideaal, goed, voldoende en slecht en vervolgens een correctie op de standaard ruwheidswaarde voor het aanwezige oever materiaal toe te passen. De invloed van begroeiing is in rekening gebracht volgens het door Ven. T. Chow ontwikkelde concept. Verschillende methoden om de equivalente ruwheid te bepalen zijn met elkaar vergeleken en de methode die door Méndez is voorgesteld blijkt de beste resultaten te geven nadat die getest werd met de gegevens van Krüger.

De methoden om het ontwerp debiet te bepalen en om kanalen te dimensioneren voor moderne irrigatie systemen, die gebaseerd zijn op het huidige concept van gewas gerelateerde irrigatie behoefte, programma's voor wateraanvoer en -verdeling aan de tertiaire eenheden zijn diepgaand bestudeerd. De gewas selectie hangt af van de bodemsoort, de waterbeschikbaarheid, de sociaal economische en klimatologische omstandigheden. Het type gewas samen met de bodemsoort bepaalt de irrigatie methode en de manier waarop het irrigatie water wordt aangevoerd, terwijl het type gewas en klimatologische omstandigheden de waterbehoefte bepalen. De vereiste hoeveelheid water in een kanaal wordt dan bepaald door de wijze waarop het water vanuit het kanaal wordt toegevoerd naar de lagere orde kanalen of de velden om aan de waterbehoefte te voldoen.

De meeste sediment transport formules gaan uit van een kanaal met een onbeperkte breedte zonder rekening te houden met het effect van de oevers op de waterstroom en het sediment transport. Het effect van de oevers op de snelheidsverdeling in zijdelingse richting is daarbij verwaarloosd en daarom wordt aangenomen dat de snelheidsverdeling en het sediment transport in elk punt van de dwarsdoorsnede constant zijn. Die veronderstelling leidt tot de aanname dat de bodemschuifspanning, de snelheidsverdeling en het sediment transport gelijkmatig verdeeld zijn. De meeste irrigatie kanalen zijn niet-breed en hebben een trapeziumvorm met uitzondering van kleine en beklede kanalen die rechthoekig kunnen zijn. In een trapeziumvormig dwarsprofiel verandert de waterdiepte en ook de schuifspanning van punt tot punt. Het effect zou meer uitgesproken zijn als de bodembreedte - waterdiepte verhouding (B-h verhouding) klein is.

Kanaalontwerp methoden en sediment transport in Nepal

Voor het ontwerp van onbeklede kanalen met sediment transport zijn in de praktijk twee benaderingen beschikbaar, namelijk de regime methode en de rationele methode. Het ontwerp volgens de regime methode bestaat uit een reeks empirische vergelijkingen die ontwikkeld zijn op basis van waarnemingen van kanalen en rivieren die een dynamisch evenwicht hebben bereikt. De rationele methode is een meer analytische benadering en daarin worden drie vergelijkingen gebruikt, namelijk een alluviale weerstandsrelatie, een vergelijking voor het sediment transport en een breedte-diepte relatie, die samen het bodemverhang, de waterdiepte en de breedte van een alluviaal kanaal bepalen voor een gegeven debiet en sediment transport en de diameter van het bodem materiaal.

De ontwerp handboeken van het Ministerie van Irrigatie in Nepal adviseren het gebruik van de regime vergelijkingen van Lacey en de tabellen van White–Bettis-Paris met de wrijvingsweerstand vergelijkingen voor het ontwerp van aarden kanalen, die ook sediment afvoeren. Maar in de praktijk is er geen consistentie in de ontwerp methoden, die zelfs van kanaal tot kanaal en zelfs binnen hetzelfde irrigatie systeem variëren. Het gebruik van de vergelijking van Lacey om de B-h verhouding te bepalen heeft over het algemeen tot bredere kanalen geleid. Dit komt omdat men de taluds flauwer ontwerpt dan uit de vergelijkingen van Lacey volgt in verband met de stabiliteit van de grond.

De tabellen van White–Bettis-Paris zijn bepaald uit hun alluviale wrijvingsvergelijkingen (1980) en uit de sediment transport vergelijkingen van Ackers en White (1973). Informatie over het gebruik van deze methode voor kanaal ontwerpen is niet gevonden en daarom kon niet worden nagegaan hoe de resultaten in termen van sediment transport zijn. Nochtans de Ackers en White vergelijkingen geven te hoge waarden voor het sediment transport wanneer die vergeleken worden met de gegevens van het Sunsari Morang Irrigatie Systeem. Het sediment dat in de kanalen binnenkomt is meestal fijn (d50 < 0.2 mm) en de meeste grote irrigatie systemen in Nepal hebben dezelfde geo-morfologische opbouw. Dit betekent dat de tabellen van White–Bettis-Paris resulteren in een kanaal met een flauwer verhang dan eigenlijk nodig is om in Sunsari Morang Irrigatie Systeem en gelijkwaardige irrigatie systemen in Nepal sediment van deze grootte te transporteren. De analyse heeft ook aangetoond dat de vergelijkingen van Brownlie en Engelund en Hansen beter bruikbaar zijn voor het type sediment dat in Sunsari Morang Irrigatie Systeem wordt aangetroffen.

Tijdens de modernisering van Sunsari Morang Irrigatie Systeem zijn de secundaire kanalen S9 en S14 volgens twee verschillende methoden ontworpen. Het secundaire kanaal S9 is ontworpen volgens het regime concept van Lacey terwijl het secundaire kanaal S14 is ontworpen met de energie benadering. In de energie benadering wordt de erosie beperkt door de wrijvingskracht niet te groot te laten worden en sedimentatie wordt beperkt door de energie in stroomafwaartse richting gelijk te houden of niet te laten afnemen. De beide kanalen zijn geëvalueerd op hun vermogen om het sediment te transporteren voor de specifieke sediment kenmerken. Het transporterend vermogen van beide kanalen (ongeveer 230 ppm) blijkt minder te zijn dan het verwachte transporterend vermogen (ongeveer 300 - 500 ppm). De energie benadering veronderstelt dat het sediment transport evenredig is met het product van snelheid en bodemverhang. Het transporterend vermogen van het kanaal dat volgens deze methode ontworpen is blijkt echter over zijn lengte veranderlijk te zijn. Dit betekent dat het sediment transport niet alleen een functie van verhang en waterdiepte is, zoals is aangenomen bij het energie concept.

Een verbeterde benadering voor het ontwerp en beheer van irrigatie kanalen

In het algemeen is de betrouwbaarheid van sediment transport formules laag en in het gunstigste geval kunnen zij slechts globale schattingen geven. Volgens Vanoni (1975) kan zelfs in de meest gunstige omstandigheden een waarschijnlijke fout in de grootte van 50-100% verwacht worden. Er is geen universeel erkende formule waarmee het sediment transport kan worden voorspeld. De meeste formules zijn gebaseerd op laboratorium gegevens met een beperkt aantal sediment en water variabelen. Daarom zouden zij moeten worden aangepast om hen geschikt te maken voor de specifieke doeleinden, anders zullen de voorspelde resultaten onrealistisch zijn. Een verbeterde rationele benadering is voor het ontwerp van alluviale kanalen met sediment transport voorgesteld. Om de bodembreedte, bodemverhang en waterdiepte van een kanaal voor een gegeven afvoer en sediment kenmerken te bepalen zijn drie vergelijkingen nodig, namelijk een sediment transport formule (totale lading), een vergelijking voor de wrijvingsweerstand (Chézy) en een relatie om de B-h verhouding te schatten.

Een kanaalontwerp programma DOCSET (Design Of Canal for SEdiment Transport) is uitgewerkt op basis van de verbeterde ontwerp methode met inbegrip van boven genoemde verbeteringen. Het programma kan ook worden gebruikt om bestaande ontwerpen voor een specifiek debiet en sediment kenmerken te evalueren. De basis eigenschappen van de nieuwe methode zijn:

– *concept maatgevende concentratie.* In plaats van het gebruik van de maximum concentratie, stelt de methode voor om een concentratie te bepalen die resulteert in een netto minimum erosie/deposito gedurende één gewas kalenderjaar;
– *bepaling van de ruwheid.* De voorgestelde methode maakt gebruik van de uitgewerkte en meer realistische bepaalde ruwheidswaarde in het ontwerp. De ruwheid van de dwarsdoorsnede wordt aangepast aan de hydraulische en sediment kenmerken. Bovendien zijn de invloeden van de oevers en de B-h verhouding inbegrepen bij de bepaling van de equivalente ruwheid van de sectie. Dit zou in een nauwkeuriger bepaling van de hydraulische en sediment transport kenmerken resulteren en dus in een beter ontwerp;
– *expliciet gebruik van de sedimentparameters.* De sediment concentratie en de representatieve diameter (dm) worden in de ontwerpfase expliciet gebruikt. Dat zal tot een flexibeler ontwerp proces leiden omdat de verschillende kanalen

verscheidene hoeveelheden sediment van diverse grootte (dm) en hoeveelheden zouden moeten kunnen afleiden en transporteren;

- *gebruik van een correctie factor.* Een correctie factor is gebruikt om de invloed van niet-brede kanalen, talud hellingen en snelheid exponenten in de sediment transport vergelijking te corrigeren. Deze aanpassing zou de nauwkeurigheid van de schatting van het sediment transport in irrigatie kanalen moeten verbeteren, een situatie waarvoor genoemde vergelijkingen niet werden afgeleid;
- *concept van een geïntegreerd systeemontwerp.* Deze benadering ziet een kanaal netwerk als één eenheid. Een kanaal systeem kan in verschillende kanalen worden onderverdeeld en een kanaal in verschillende secties, afhankelijk van het debiet. Maar het beheerplan voor het water en het sediment wordt vooraf voor het hele kanaalnetwerk opgesteld. Daarna kan een kanaal op basis van dit beheerplan hydraulisch worden ontworpen;
- *Selectie van de B-h verhouding.* Een criterium voor de selectie van de B-h verhouding is vastgesteld nadat de criteria voor de selectie van de talud helling in Nepal en de specifieke kenmerken van het sediment transport zijn bestudeerd.

Aangezien het sediment transport beïnvloed wordt door het beheer van het irrigatie systeem, zou het ontwerp zich moeten richten op een kanaal dat voldoende flexibel is om aan de vraag naar water te voldoen en toch tot een minimum aan sedimentatie/erosie leidt. Voldoende transport capaciteit tot aan de gewenste plaats (doorvoer), voorzieningen voor gecontroleerde sedimentatie wanneer de beheerplannen de transportcapaciteit beperken (zandvangen) en voorbereiding van onderhoudsplannen (bagger werkzaamheden) zijn enkele aspecten die in een ontwerp moeten worden geanalyseerd en meegenomen om de sediment transport problemen te verminderen.

Kanaal ontwerpen kunnen alleen de beste kanaal afmetingen voor een specifiek debiet en sediment concentratie geven. Voor debieten en sediment concentraties buiten de ontwerpwaarden kan erosie of sedimentatie optreden. Een ontwerp zou daarom een evenwicht moeten vinden tussen de totale erosie en sedimentatie gedurende één groeiseizoen. Daarom moet een ontwerp niet uitgaan van de maximum sediment concentratie die tijdens het irrigatie seizoen wordt verwacht, maar van een waarde die in een minimale netto erosie/sedimentatie resulteert. De beste manier om een kanaal onder dergelijke scenario's te evalueren is de toepassing van een geschikt sediment transport model. Bovendien de ruwheid van een kanaal hangt af van hydraulische condities, sediment kenmerken en onderhoud plannen die tijdens het irrigatie seizoen continue veranderen. De kanalen worden ontworpen op basis van de aanname van een eenparige stroom en een sediment transport dat in evenwicht is, deze voorwaarden worden echter zelden in irrigatie kanalen aangetroffen omdat de wateraanvoer continue geregeld wordt om aan de wisselende waterbehoefte te voldoen. Vandaar dat een kanaal ontwerp moet worden geëvalueerd met behulp van een sediment transport model om de juiste ontwerp criteria te selecteren en het ontwerp op basis van de voorgestelde waterbeheer plannen te evalueren.

Het computer model SETRIC

Het computer model SETRIC is een eendimensionaal model, dat de waterstroom in een kanaal schematiseert als een quasi-permanente, geleidelijk veranderende stroom. De eendimensionale stromings vergelijking wordt opgelost met behulp van de predictor-corrector methode. Het over de diepte geïntegreerde model van Galappatti is gebruikt om de werkelijke sediment concentratie in elk punt van het dwarsprofiel onder niet-evenwichts omstandigheden te bepalen. Het model van Galappatti is gebaseerd op de 2-D convectie-diffusie vergelijking. De massa vergelijking voor het totale sediment transport is opgelost met behulp van de aangepaste Lax methode, waarbij aangenomen is dat de sediment concentratie in een permanente toestand verkeert. Voor de bepaling van de evenwicht concentratie kan één van de drie formules voor totaal sediment transport, namelijk Brownlie, Engelund en Hansen of Ackers en White, gebruikt worden.

Het model SETRIC is geëvalueerd met behulp van andere hydrodynamische en sediment transport modellen (DUFLOW en SOBEK-RIVIER) en werd gevalideerd met veldgegevens van SMIS. De nauwkeurigheid van de verschillende sediment transport vergelijkingen is vergeleken. De methoden van Brownlie en Engelund en Hansen zijn redelijk nauwkeurig voor een sediment diameter van 0.1 mm (d50), terwijl de voorspelbaarheid van Ackers en White voor deze sediment grootte minder goed is. De gevoeligheid van de methode van Brownlie was uniformer dan de andere twee methoden voor sediment grootte van 0.05 tot 0.5 mm.

Veldonderzoek

Voor het veldonderzoek naar het sediment transport gedrag is het secundaire kanaal van SMIS (S9) geselecteerd. Omdat het doel van het veldonderzoek het testen van de ontwerp methode voor kanalen met sediment transport was; werd de voorkeur aan een onlangs ontworpen en aangelegd kanaal gegeven. Het veldonderzoek naar de irrigatie aspecten en het sediment transport werd in 2004 en 2005 uitgevoerd. Tijdens het veldonderzoek werd het debiet in het secondaire kanaal S9 gemeten. Een stuw met een brede kruin onmiddellijk stroomafwaarts van de inlaat is gekalibreerd en gebruikt voor de debietmetingen. Voor de bepaling van de sedimentconcentratie, werd dagelijks punt bemonstering door onderdompeling van een fles en benedenstrooms van de watersprong uitgevoerd. De monsters werden naar het laboratorium gebracht om de sediment concentratie te bepalen. Puntbemonstering over het dwarsprofiel met behulp van een pomp werd ook uitgevoerd en toonde aan dat punt bemonstering ongeveer 8% lagere waarden voor al het materiaal in suspensie en ongeveer 35% lagere waarden voor het materiaal in suspensie groter dan 63 μm gaf. Aan het eind van het irrigatie seizoen is het bodem sediment langs het kanaal bemonsterd om de sediment grootte en andere eigenschappen te bepalen.

Het irrigatie debiet dat aan de subsecundaire kanalen wordt geleverd, de wateraanvoer schema's en de waterpeilen stroomopwaarts van de kunstwerken werden ook gemeten. Met het oog op de morfologische veranderingen zijn de kanaal afmetingen voor en na het irrigatie seizoen opgemeten. De snelheidsverdeling in de trapeziumvormige dwarsdoorsnede van de aarden kanalen is ook gemeten. Bovendien is de wandruwheid (indirecte meting) ook aan het begin, midden en eind van de seizoenen bepaald om de veranderingen in de ruwheid met de tijd te bepalen.

Resultaten van het modelonderzoek

Het model SETRIC is gebruikt om het effect van beheer activiteiten op het sediment transport te bestuderen, vooral de verandering in de vraag naar en levering van water, de wijze van wateraanvoer die gebaseerd is op het beschikbare water en de verandering in sediment concentratie door variatie in sediment aanvoer in de rivier of problemen met de juiste bediening van de zandvang. Voor de wateraanvoer naar het secondaire kanaal S9 is een beheerprogramma ontwikkeld en geëvalueerd met het oog op de efficiency van het sediment transport onder veranderende omstandigheden van de sedimentaanvoer. De verbeterde methode voor het kanaal ontwerp is geëvalueerd door de resultaten te vergelijken met de resultaten voor het bestaande ontwerp van het secondaire kanaal S9. De belangrijkste bevindingen uit het onderzoek met het model SETRIC zijn:

- de beheerprogramma's voor de watertoevoer kunnen zodanig ontworpen en uitgevoerd worden dat erosie/sedimentatie problemen in bepaalde kanaalsecties verminderen, zelfs nadat het systeem is opgeleverd en gebruikt wordt;
- het ontwerp van en aannamen voor de beheerplannen zijn voor het secondaire kanaal S9 niet opgevolgd. Vanuit het sediment transport perspectief brengt de bestaande beheerpraktijk meer schade toe aan de subsecundaire en tertiaire kanalen dan aan het secondaire kanaal S9;
- de periodieke veranderingen in de waterbehoefte en de daaraan gekoppelde veranderingen in de sedimentaanvoer kunnen zodanig beïnvloed worden dat er een evenwicht in het seizoensgebonden erosie/sedimentatie proces optreedt. Tijdens een periode, kan er sedimentatie optreden maar dat kan tijdens de volgende periode worden geërodeerd;
- het voorgestelde beheerplan voor de watertoevoer is gebaseerd op het bestaande kanaal en zijn regelkunstwerken, en geldt voor debiet schommelingen van ongeveer 46% tot 114%. Daarom kan het plan voor de huidige infrastructuur worden ingevoerd en kan het alle mogelijke wateraanvoer situaties in het kanaal aan;
- het voorgestelde wateraanvoer plan garandeert of de volledige of geen watertoevoer naar de subsecundaire kanalen die volgens hetzelfde principe ontworpen zijn. Dit plan zou het bestaande sedimentatie probleem in deze kanalen moeten verminderen;
- een juist beheer van de zandvang is essentieel voor de duurzaamheid van het Sunsari Morang Irrigatie Systeem;
- de secundaire kanalen zouden water in rotatie moeten krijgen wanneer de vraag afneemt of minder water in het hoofdkanaal beschikbaar is. Deze rotatie zorgt ervoor dat het ontwerpdebiet in de secundaire kanalen stroomt en daardoor zal het sedimentatie probleem verminderen. Het hoofdkanaal zou vanuit het oogpunt van sediment transport en wateraanvoer moeten worden geanalyseerd om te bepalen welke rotatiemethode de beste is.

Belangrijke bijdragen van dit onderzoek

Naast de aanbevelingen voor het ontwerp, beheer en onderhoud van het secondaire kanaal S9 die vanuit een sediment transport perspectief kunnen worden gemaakt, kunnen ook de volgende bijdragen van dit onderzoek genoemd worden:

- de snelheid- en schuifspanning verdeling over het trapeziumvormige kanaalprofiel zijn gedetailleerd bestudeerd om de correctie factor voor de sediment transport formules af te leiden. Dit zal de nauwkeurigheid van deze formules en de analyse van het sediment transport in irrigatie kanalen verbeteren;
- een expliciete methode om de ruwheidparameters te bepalen die nodig zijn voor de berekening van de equivalente ruwheid in het computer model;

- het sediment transport model SETRIC is verbeterd. Het model is verder verbeterd en kan nu zowel als ontwerp- en als onderzoek hulpmiddel worden gebruikt voor de analyse van sediment transport onder verschillende beheerprogramma's voor irrigatie systemen;
- een verbeterde methode voor het ontwerp en beheer van irrigatie kanalen is uitgewerkt en gepresenteerd. Een nieuw computerprogramma DOCSET is gebaseerd op de verbeterde benadering. Het programma is interactief, eenvoudig te gebruiken, ook door ontwerpers met een beperkte kennis van modelleren;
- een beheerplan voor de wateraanvoer is uitgewerkt en getest met het oog op de veranderende aanvoer van water en sediment en dat met de bestaande infrastructuur kan worden uitgevoerd;
- de oorzaken van sedimentatie in de subsecundaire kanalen van S9 zijn vastgesteld en benoemd.

Conclusies en vooruitzichten voor de toekomst

Het kanaalontwerp is een iteratief proces waarbij het uitgangspunt de voorbereiding van beheerplannen is. Vervolgens worden de ontwerpparameters geselecteerd en wordt een voorlopig hydraulische kanaalontwerp gemaakt. De resultaten van dit voorlopige ontwerp worden dan in het model gebruikt om de voorgestelde beheerplannen en het sediment transport in het systeem te simuleren en te evalueren. De nodige aanpassingen worden, indien nodig of in de ontwerp parameters of in de beheerplannen gemaakt. Dan wordt het kanaal opnieuw ontworpen en het proces zou moeten worden voortgezet tot een bevredigend resultaat is verkregen.

De grovere fractie van het sediment wordt meestal uitgefilterd bij de hoofdinlaat en de zandvang van een irrigatie systeem. Het sediment dat in het hoofd- en secundair kanaal getransporteerd wordt is over het algemeen fijn zand. Het grootste deel van de slibfractie (sediment < 63 μm) wordt vervoerd naar de subsecundaire en tertiaire kanalen waar het neerslaat. Ook is vastgesteld dat dit fijne sediment niet naar de kanaalbodem rolt zoals dat voor zand normaal is, maar op het talud wordt gedeponeerd. Op deze manier wordt het kanaalprofiel smaller en het talud steiler. Dit kan niet verklaard worden door de huidige aannamen voor sediment transport en nader onderzoek van dit aspect zou gericht moeten zijn op het transportproces van fijn sediment om het ontwerp en beheer van irrigatie kanalen te verbeteren.

De flexibiliteit van beheer en sediment transport aspecten beperken elkaar. Een kanaal zonder enige regeling kan wat betreft sediment transport met meer betrouwbaarheid ontworpen en beheerd worden. Zodra de waterstroom geregeld wordt, verandert het sediment transport patroon en het ontworpen kanaal zal zich anders gaan gedragen. Vandaar dat zowel flexibiliteit als een efficiënt sedimentbeheer moeilijk tegelijkertijd te verwezenlijken zijn. Een compromis is nodig en dit zou in het ontwerp moeten worden teruggezien.

Alle methoden om het sediment te transporteren, uit te sluiten of neer te laten slaan zijn tijdelijke maatregelen en verplaatsen het probleem slechts van één plaats naar een andere. Zij geven niet een allesomvattende oplossing voor het sedimentprobleem. Een beter begrip van sediment transport zal nodig zijn om de problemen vooraf te identificeren en de best mogelijke oplossingen te zoeken.

About the author

Krishna P. Paudel graduated in civil engineering with honours from the University of Roorkee (now IIT Roorkee), India in 1989. In February 1990, he joined the Department of Irrigation under the Ministry of Water Resources. From 1990 till 2003, he worked at different district irrigation offices and large scale irrigation schemes. During that period he prepared detailed designs and feasibility reports of more than 40 irrigation schemes and constructed/rehabilitated 32 irrigation schemes (60 ha to 1200 ha command area). His main responsibility at the district irrigation offices was planning, design and implementation of irrigation development and flood protection related works in the district. The major works performed were: collection and analysis of hydrological and meteorological data; preparation of feasibility assessment reports, design of the schemes and preparation of bidding documents, institutional development work of Water Users Associations, construction management and quality control, training of the farmers for scheme operation and maintenance, financial management and coordination with other district agencies working in the field of water and agriculture.

From 1994 to 1997, he worked in Chandra Canal System (10,000 ha). His responsibility was to operate and maintain the irrigation scheme. He also prepared the detailed design and inventory of the Chandra Canal System for scheme rehabilitation. He worked in the Sunsari Morang Irrigation Scheme (68,000 ha) from 1998 to 2000 and was deputed at the headworks to operate and maintain them, the settling basin and the main canal.

In October 2000, he joined IHE (now UNESCO-IHE), Delft, the Netherlands and acquired his MSc Degree in Hydraulic Engineering specialization in Land and Water Development with Distinction in 2002. After his MSc, he resumed to his duty in the Department of Irrigation and was involved in irrigation development related works of the district irrigation offices.

In January 2004, he joined the Core Land and Water Development of UNESCO-IHE, the Netherlands as a PhD research fellow. His thesis entitled: "Role of Sediment in the design and management of irrigation canals: Sunsari Morang Irrigation Scheme, Nepal" has made an assessment of existing canal design methods from sediment transport perspective. The thesis analysed the factors that influence the sediment transport process in irrigation canals and proposed a design approach that suggests integrating the hydraulic design, management and modelling aspects to prepare a canal design. He developed the canal design program DOCSET that can be used to design and evaluate irrigation canals for sediment transport. During the research he has improved, verified and used the sediment transport model SETRIC for analysis of the scheme.

In May 2008, Mr. Paudel joined CMS Nepal (Development Consultants) as Senior Water Resources Engineer and has been involved as Construction Management Engineer for the construction of the headworks and settling basin of Sikta Irrigation Project (34,000 ha).

Mr. Paudel presented papers for the International Commission on Irrigation and Drainage (ICID) in Beijing in 2005. He has published one and submitted two papers in international peer reviewed journals.

T - #0079 - 071024 - C296 - 246/174/16 - PB - 9780415615792 - Gloss Lamination